THE ORIGINS OF
CREATIVITY

Journey to the Ants: A Story of Scientific Exploration,
with Bert Hölldobler (1994)

Naturalist (1994; new edition, 2006)

The Diversity of Life (1992)

The Ants, with Bert Hölldobler (1990);
Pulitzer Prize, General Nonfiction, 1991

Success and Dominance in Ecosystems:
The Case of the Social Insects (1990)

Biophilia (1984)

Promethean Fire: Reflections on the Origin of the Mind,
with Charles J. Lumsden (1983)

Genes, Mind, and Culture, with Charles J. Lumsden (1981)

On Human Nature (1978);
Pulitzer Prize, General Nonfiction, 1979

Caste and Ecology in the Social Insects, with George F. Oster (1978)

Sociobiology: The New Synthesis (1975; new edition, 2000)

The Insect Societies (1971)

A Primer of Population Biology, with William H. Bossert (1971)

The Theory of Island Biogeography,
with Robert H. MacArthur (1967; reprinted 2001)

Edward O. Wilson

THE ORIGINS OF CREATIVITY

ALLEN LANE
an imprint of
PENGUIN BOOKS

ALLEN LANE

UK | USA | Canada | Ireland | Australia
India | New Zealand | South Africa

Allen Lane is part of the Penguin Random House group of companies
whose addresses can be found at global.penguinrandomhouse.com.

First published in the United States of America by W. W. Norton & Company, Inc. 2017
First published in Great Britain by Allen Lane 2017
001

Copyright © Edward O. Wilson, 2017

The moral right of the author has been asserted

Frontispiece: The Reality Unseen where science and the humanities meet
(Calley O'Neill and Rama the Elephant. Painting from
The Rama Exhibition, A Journey of Art and Soul for the Earth.
www.TheRamaExhibition.org. Used by permission).

Printed in Great Britain by Clays Ltd, St Ives plc

A CIP catalogue record for this book is available from the British Library

ISBN: 978–0–241–30920–9

CONTENTS

I

II

III

IV

V

The term "humanities" includes, but is not limited to, the study and interpretation of the following: language, both modern and classical; linguistics; literature; history; jurisprudence; philosophy; archaeology; comparative religion; ethics; the history, criticism and theory of the arts; those aspects of social sciences which have humanistic content and employ humanistic methods; and the study and application of the humanities to the human environment with particular attention to reflecting our diverse heritage, traditions, and history and to the relevance of the humanities to the current conditions of national life.

—*National Foundation on the Arts and the Humanities Act, U.S.A., 1965, as amended*

A wolf defeated by a lion. A fable of fallen pride. (Benjamin Carlson, *The Wolf and His Shadow*, 2015. Ink on Illustration Board. 20 x 30 inches. © Benjamin Carlson. Benjamin Carlson's *The Wolf and his Shadow* [p. 44 in *Call of the Wild* 2015–16 Aesop's Fables, by Bronwyn Minton, Associate Curator of Art and Research, The National Museum of Wildlife Art].)

I

The humanities arose from symbolic language, a capacity that singly and dramatically distinguishes our species from all others. Coevolving with the structure of the brain, language freed the mind from the animal to be creative, thence to enter and imagine other worlds infinite in time and space. We were empowered but, as I will show in this first part, we retained the emotions of our ancient primate ancestors. The combination, composing what we roughly call the humanities, is why we are both supremely advanced, and supremely dangerous.

1

THE REACH OF CREATIVITY

Creativity is the unique and defining trait of our species; and its ultimate goal, self-understanding: What we are, how we came to be, and what destiny, if any, will determine our future historical trajectory.

What, then, is creativity? It is the innate quest for originality. The driving force is humanity's instinctive love of novelty—the discovery of new entities and processes, the solving of old challenges and disclosure of new ones, the aesthetic surprise of unanticipated facts and theories, the pleasure of new faces, the thrill of new worlds. We judge creativity by the magnitude of the emotional response it evokes. We follow it inward, toward the greatest depths of our shared minds, and outward, to imagine reality across the universe. Goals achieved lead to further goals, and the quest never ends.

The two great branches of learning, science and the

humanities, are complementary in our pursuit of creativity. They share the same roots of innovative endeavor. The realm of science is everything possible in the universe; the realm of the humanities is everything conceivable to the human mind.

Drawing on the combined consciousness of our species, each one of us can go anywhere in the universe, seize any power, achieve any goal, search for infinity in space and time. Of course, it is also true that when ruled by wild surmise and the animal passions we all share, our unbounded fantasy can disintegrate into madness. Milton expressed this greatest risk of the human condition very well.

A mind is its own place, and in itself
Can make a Heav'n of Hell, and a Hell of Heav'n.

So perhaps it is a blessing that the mind does not travel easily to vast and unfamiliar regions, but instead prefers to cruise and repeat its excursions in small familiar circles. Further, as a rule, people do not like the solitude of their own thoughts. A team of psychologists from the University of Virginia and Harvard University recently found that volunteers disliked sitting alone for even as little as six minutes with nothing to do but think. They enjoyed mundane external activities more. They even preferred administering electric shocks to themselves if nothing else was available.

The full explanation of any biological phenomenon, including creativity in both science and the humanities, engages three levels of thought. First, for any conceivable living entity or process—a bird taking flight, a lily growing toward the sun, your reading of this sentence—the first inquiry must be, *What* is it? Provide the structure and functions that define the phenomenon. If it involves music or theater, perform it. The second level is the question, *How* was it put together? What made it come into existence? What were the events that resulted in the conditions of its origin, whether ten seconds or a thousand years ago? The third and final level is, *Why* do the phenomenon and its preconditions exist in the first place? Why not a different mode of evolution not present on this planet that might have produced a different kind of thinking brain?

Scientists study living phenomena at all three of these levels. As a rule, they choose entities and processes that engage the what, how, and why in whatever detail and dimension that lies within their reach.

Biologists, however, perhaps even more than other scientists, feel it necessary to seek cause and effect at all three levels. The causes that bring about a living phenomenon, such as the flight of a bird or our perception of a flower's colors, are called *proximate causes*. The events that guided the evolution of the phenomenon to its present state are called *ultimate causes*. Proximate causes are the *what* and *how* of a full explanation. Ultimate causes are the *why*.

Scientific explanations of organic life, including human life, routinely entail both proximate and ultimate causes. In contrast, accounts that govern inquires in the humanities attempt at best only proximate explanations. Ultimate causation tends to be left to the God of Genesis, or ancient alien visitors, or a *mysterium tremendum et fascinans* imagined to be resident deep within the human mind. Take as a random example the red color of a flower petal placed before you. The redness, like all other hues, is distinguished by the stimulation of a particular portion of the electromagnetic spectrum composing the visual spectrum. The receptors are red–sensitive cones in the retina. The latter transmit signals to a staging center in the brain cortex, thence in relays back to the rear of the cortex, then to integrating units of perception and emotion, and finally back out to the centers of the conscious forebrain, causing you to say "red" (or perhaps *rot*, or *rouge*, or *krasnyy*, or *bombu*, depending on your native language).

Having gone this far, scientists have in recent decades tracked down the interacting segments of DNA that compose the genes that prescribe the recognition of colors.

Thus, scientific research has brought us close to solving the first line of mysteries in human color vision. Yet the way is now open to the still deeper question of ultimate causation: Why can humans see a particular spectrum of color but cannot see infrared, or ultraviolet, or other frequencies outside the narrow segment of the electromagnetic spectrum yielding visible light? And even deeper, why does DNA, and not some other coding chemical, prescribe color vision and all the other processes of life on Earth? Might we expect to find fundamentally different codes on life-bearing exoplanets? And why do we see color in the first place, instead of just shades of light and dark?

The answers of such *why* questions await the reconstruction of prehistory, during which our species evolved from earlier hominins, and much further back over tens of millions of years when basic properties of our present brain and senses were shaped within the earliest ancestral primates.

Scholars in the humanities have traditionally confined themselves to the *what*. They have touched lightly on the *how*. They have seldom ventured into the world of *why*. They build upon the biological particularities of the senses and emotions already in place at the dawn of the Neolithic, some ten thousand years ago—hence the almost exclusive contemporary content of the humanities: the creative arts, linguistics, history, jurisprudence, philosophy, moral reasoning, and theology.

It might seem—*feel* is perhaps the better word—that

the human suite of intellect and emotion is the only one that could have attained creativity. This diagnostic trait of our species, almost four billion years in the making, might seem to require some unique feature of evolution or else the hand of God extended special to our lineage.

That supposition, which has dominated religious thinking for millennia, would almost certainly be wrong. It is easy to find in nature alternative springboards in advanced social organization, some of which in time might have swerved in evolution to the human level. Consider the remarkable mound-building termites, technically called macrotermitines, of Africa and South America. Their earthen nests, multistoried, built of soil and feces, teem with populations of hundreds of thousands to millions, and rise in places to higher than a human head. Like those of humans, their abodes are exquisitely well designed. In some species, the nests are air-conditioned by elaborate systems of ducts that continuously circulate fresh air from the surrounding ground surface and used air, by heat convection, from the massed inhabitants up and out of the mound. Each macrotermitine colony contains a proletariat of sterile workers and their two parents, a royal pair responsible for all reproduction. How is this possible? The massive queen, the size of your two thumbs, lays a continuous stream of tiny eggs. The workers divide labor by division into physical castes, including a large army of suicidally ferocious, large-headed soldiers. (In Surinam, I once required assistance to dissect their scimitar-shaped mandibles out of the web of my right thumb.)

The inhabitants remain underground within the mound and the meters-deep labyrinth of galleries and chambers excavated below the mound, with a couple of notable exceptions: the flights of virgin queens and their consort males that establish new colonies, and swarms of workers that come out at night to search for bits of dead vegetation. Placing an ear close to the nests (not too close!), one hears a faint hiss raised by countless tiny footfalls. The nocturnal harvests are used to grow an edible fungus in subterranean gardens.

The macrotermitines are true superorganisms. The collective intelligence of each of the colonies is still far below the level of humans and other mammals, even below most birds, but well above that of solitary insects. Yet their creativity remains zero. Let us suppose they had ascended during their evolution to the human level. Their "termiteries," if I may coin a term, would include the following: a love of absolute darkness (and a Dracula-like panic at the merest hint of daylight); an exclusive diet of cultivated mushrooms; sex limited to royals; and death to all would-be immigrants, even from the same species. The sick and the injured of the colony promptly and peremptorily eaten; no hospitals, no pity.

Consider this. In a century or two, space technology may well provide a first close look at exoplanets, the planets of other star systems. An intensive search for evidence of life will certainly follow. If life is found, and one or more intelligent species, we should be prepared for, well, anything.

THE BIRTH OF THE HUMANITIES

The humanities were not born of the oral epics of Mycenaean Greece, early Sumerian script pressed into clay, or the graven gods of predynastic Egypt. These rudiments are all younger than ten millennia, only an eyeblink in the history of our species. Nor will we find the dawn of the humanities in the cave paintings of Chavet and Sulawesi, or the bird-bone flutes of the Swabian Jura. These artifacts, oldest known of their kind, were fashioned a little more than thirty millennia ago.

The birth of the humanities occurred much farther back in time, closer to one thousand millennia, and it took place in the site and circumstance that upon reflection seem the most logical: the nocturnal firelight of the earliest human encampments.

Such is the reconstruction being pieced together by a diverse array of researchers in paleontology, anthropology, psychology, brain science, and evolutionary biology. The research is part of the search for the origin of the human species itself, a grail of both science and the humanities. It is a reminder that history is the story of cultural evolution, while prehistory is that of genetic evolution. Prehistory tells us not only what happened prior to cultural history, but also why the human species as a whole followed one particular trajectory and not another.

Let us peer back in deep time for a moment. In order to compare various trajectories that could have been followed, researchers have at their disposal a large number of living Old World primates. Among these subjects are the monkeys and apes, our closest surviving phylogenetic cousins. They in turn include stages close to what most likely were the human precondition.

I've had the pleasure as a biologist focused on social evolution to observe two species in the wild: vervets and baboons. I've also spent three days in the midst of a semi-wild population of rhesus macaques, tutored by the pioneering primatologist Stuart Altmann. I've had lunch at the Yerkes Primate Research Center at Emory University with Kanzi, an intensely studied bonobo raised in captivity. Kanzi's species, which used to be called pygmy chimpanzees, are considered the most humanlike of all the primates.

Expert studies of these and similar species have revealed

that most of their time is spent in exploring their environment for food. A smaller proportion is devoted to social interactions, including bonding, dominance and submission, grooming, courtship, mating, handling of infants, food discovery and retrieval, leadership, and conformity.

Studies by experts on each of the Old World monkeys and apes have increasingly focused on what each group member learns while it watches and interacts with other members. The center of interest is how the animal uses this social information. When it imitates a groupmate, the scientists ask, what exactly is it imitating—the exact movements of the groupmate, or the consequence of its movements? If, for example, the groupmate pulls back a clump of grass to retrieve grasshoppers for a snack, what will the observer learn from this action—that turned grass yields food, or, alternatively, that the act of turning the grass yields food?

Inasmuch as the individual knows each of its groupmates as individuals, and can understand and predict their behavior, and further understands what the consequences will likely be, it can use this knowledge to its own personal advantage. Most importantly for the group, the observing animal knows how, when, and whether to compete or to cooperate. The informed interplay between competition and cooperation is the flywheel of a successful social organization.

Social primates, from langurs and macaques at the low end to chimpanzees and bonobos at the high end, have brains large enough to perceive the mood and likely response of a groupmate to a variety of situations. A member of such a well-organized society knows its place and responds accordingly and accurately from one exchange to the next. The most successful member of a stable society has a strong sense of empathy. It can see what others see, feel what they feel, and gauge its response with precision— when to advance and when to retire, whom to groom and whom to avoid, whom to challenge and whom to placate.

Empathy, the intelligence to read the feelings of others and predict their actions, is not the same as sympathy, the emotional concern felt for another's plight combined with a desire to provide help and relief. It is closely related to sympathy, however, and led to sympathy in the course of human evolution.

It follows that the best way for a scientist to study social behavior is to enter their lives with deliberate empathy and sympathy, to know them individually in as much detail and as close as possible. Frans de Waal, the foremost expert on chimpanzee social behavior, states his investigative credo as follows.

My profession depends on being in tune with animals. It would be terribly boring to watch them for hours without any identification, any intuition about what is going on, any ups and downs related to their ups and downs. Empathy is my bread and butter, and I have made many a discovery by closely following the lives of animals and trying to understand why they act the way they do. This requires that I get under their skin. I have no trouble doing so, love and respect animals, and do believe that this makes me a better student of their behavior.

Social animals at or close to our level are hardwired to be that way. Neurobiologists have come to recognize the existence of three neural routes activated in the brains of human and other advanced primates during social interactions. The first is mentalizing, in which goals are formed and appropriate activities planned to meet them. The second is empathizing, putting oneself in the skin of another to access their motives and feelings and anticipate their future actions. Empathy is a kind of gaming, through which the individual communicates with the group and the group thereby organizes itself.

Finally, there is mirroring, by which the individual senses the mood and emotions of another, and experiences them to some degree. Mirroring leads readily to imitation of successful strategies by others. It is also part of the gateway to sympathy and, among human beings at least, what we have come to treasure as the quality of mercy.

Empathy and mirroring have evidently evolved to a degree that corresponds to the average amount of time members of the group interact with one another. Measurements of time show that such a correlation exists. Free-ranging savanna baboons (*Papio cyanocephalus*) spend less than 10 percent of their time socializing and 60 percent foraging and feeding. Vervet monkeys (*Cercopithecus aethiops*) spend 40 percent of their time foraging and feeding, and less time than baboons in socializing.

Human beings spend a great deal more of their time socializing than do these and other Old World primates. Even though their schedules vary enormously according to occupation, people have always tended to meet in groups and engage in social exchanges between intervals of solitude. In the developed world, social life has been expanded almost endlessly through public entertainment and social media.

Was human gregariousness the Darwinian driving force that led to our high levels of social intelligence—in particular, empathy, mirroring, and problem solving? Yes, but gregariousness was only part of the evolutionary process that created the human condition. For the complete story, we need to turn to the unique origin of social behavior in the ancestral hominins to the present time, as it has been construed by experts. The signal event was a massive increase in brain size, mostly entailing the frontal lobe. Starting about three million years ago, the cranial capacity of our prehuman ancestors grew from that close to chimpanzees at 400 cubic centimeters (cc) to 600 cc in the habilines (*Homo habilis*), then by a million years ago to 900 cc in our direct ancestors *Homo erectus*, and finally by the modern level (around 1,300 cc) in *Homo sapiens*.

In the process of evolution by natural selection, just as in everyday life, small events can have large, even great consequences. The small event in prehuman evolution appears to have been a shift from a primarily vegetarian diet—fruit, seeds, soft foliage—to one with substantially more meat. The shift was made easier by the habitat in which it occurred. The African savanna is a vast expanse of grassland sprinkled with riverine forests and copses of tropical trees. The harvesting of meat was made easier by the presence of animals easily tracked (for those who know how) across the open plains. It was further helped by the frequent occurrence of lightning-struck fires, which trapped and killed many of the fleeing animals. The fires also cooked some of the victims, furnishing high-energy food, rich in protein and fat and easily chewed.

As the change occurred, it necessitated an alteration of the entire gastrointestinal system, from mouth to anus. It also pushed the australopith ancestors into becoming more social. Whereas the vegetarian apes and monkeys tend to search and feed independently, it became necessary for our ancestors to cooperate more closely during forays. Then, when a large food item was found by scavenging or the downing of prey, there had to be sharing in a way that avoided potentially lethal combat. The hunting and scavenging of large animals as opposed to collecting vegetable materials further required a communal rendezvous, or den, or both.

Finally, to cap this adaptive shift (as evolutionary biologists call it), the advantage of carnivory was strengthened by the control of fire. From a nearby ground fire, it is easy to obtain burning limbs and branches and carry them to a campsite. I did it myself as a Boy Scout on the edge of a dying fire in the pine savanna of Alabama. I found that where camp fires carelessly tended can start forest fires, the reverse is also true: fire can be harvested and brought to camp. There was really no need for prehumans to start up fires by sparking flammable material with flints or spinning wood.

It is widely believed by experts that the habiline ancestors of modern humanity followed this scenario of carnivory and thereby a radical evolution of brain size and social intelligence. The theory has not yet been proven conclusively, but is supported by the fossil evidence of campsites and controlled fire in *Homo erectus*, a descendant of *Homo habilis* and direct ancestor of our own species.

Let's now go farther back into still deeper genetic history. About six million years ago, an anthropoid ape species inhabiting the African savanna split into two species. One led to the chimpanzee line, which was destined later to split again into the two modern species, the "common" chimpanzee and its more humanlike cousin the bonobo. The other line of the original divarication, proceeding through the evolutionary labyrinth of australopith species and then multiple species of *Homo*, yielded modern humanity, comprising the single, destined-to-be apocalyptic species *Homo sapiens*.

Because of the chimpanzees' close genetic relationship to humans—we share over 98 percent of our genes by common descent—scientists have studied these apes intensely for what they might reveal about the origin of the human mind and social intelligence.

Individual chimpanzees, the scientists have found, know other group members thoroughly. They behave according to their own rank and relationship among the groupmates. They have surprisingly high IQs. They can learn sequences of numbers, for example "64136 . . . ," faster and retain them better than humans. Because free-ranging chimps spend half their time in trees, this ability may well be an adaptation for quick recall and assessment of limbs and trunks, and the best sequence that can support the chimps' weight. Their arithmeticity must also have been useful in following worn paths along the ground in an

environment crowded with large predators. Lions, crocodiles, and, above all, deadly tree-climbing leopards—each
masters of ambush with their own techniques—lay almost
everywhere hidden and waiting.

Yet, although highly intelligent in at least one aspect,
chimps score far below humans in others. They live in
the here and now. They cannot plan their actions for even
the next day, whereas humans can construct scenarios
thousands of years into the future and millions of miles
into space. When supplied with paint and brush, chimps
can draw pictures, but not as well as humans of any age
beyond infancy. For example, chimps can spontaneously
draw the outline of a face, but not details within it, a feat
any young human child performs effortlessly.

Chimpanzees also fall short in the ability to cooperate
or act altruistically. Brian Hare and Jingzhi Tan, neuro-
scientists at Duke University, have summarized the evidence for both chimps and bonobos. They observed that
while the apes easily cooperate with their groupmates in
mutually beneficial behavior, they can do so only for a
few, relatively simple tasks.

We suspect that it is not a tendency to act altruistically
that makes humans unique. Instead, it seems more likely
our species is unusually cooperative because of our flex-
ible ability to avoid high-cost helping (i.e., harmful to

reproductive success) while recognizing the benefit of mutualistic endeavors . . .

To express this increasingly complex subject as succinctly as possible, the ancestors of our species developed the brain power to connect with other minds and to conceive unlimited time, distance, and potential outcomes. This infinite reach of imagination, put quite simply, is what made us great.

Psychologists and neurobiologists have taken us this far into the *what* and *how* of the human achievement. We need to press on to complete if possible the ultimate explanation of our origin. *Why* did it happen? *Why* are there humans in the first place? We know, or believe we know, how a partially carnivorous diet brought the prehuman groups to campsites and an increase in empathy, imitative capability, and cooperation. But *why* did these changes lead to a tripling of brain size, the most rapid evolutionary growth of a complex organ of all time?

The answer, at least a few anthropologists have come to believe, is actually in full sight. It has already been provided by the hunter-gatherer societies still among us around the world. The forming of campsite and control of fire brings the group together tightly in the long evening hours before sleep. They neither hunt nor gather, nor have any other reason to venture out into the surrounding darkness. They draw close and communicate where there is no other choice. This period in the daily cycle is the time to tell stories, raise status, tighten alliances, and settle scores. The fire is the life-giver. It warms and feeds the people. It creates a sanctuary of light, around which nocturnal predators circle but dare not enter. Firelight is the Prometheus that shone upon the gods and brought humanity closer to them.

For our present self-understanding, it is of consuming importance to estimate what the ancestral humans said and did in the firelight. A recent thorough record was made by the anthropologist Polly W. Wiessner of the talk around the campfire of the most famous of Earth's hunter-gatherers, the Ju/'hoansi (!Kung Bushmen) of the Kalahari Desert. Wiessner found differences between "daytime talk" and "firelight talk" that were even more striking than imagined previously. Daytime talk is focused on practical aspects of travel and the search for food and water. People working together talk about the food they seek. They also gossip back and forth in a manner that helps to stabilize their social networks. The subject matter is highly personal. Given the stringent quality of Ju/'hoansi existence, their talk is imbued with life-and-death choices. The conversation is also practical. It doesn't stray far, or play on the imagination and fantasies that are possible in periods of leisure.

In the evening the mood relaxes. In the chiaroscuro firelight the talk turns to storytelling, which drifts easily into singing, dancing, and religious ceremonies. Storytelling, especially among the men, turns frequently to successful hunts and epic adventures, their dominant daytime activity. As described by Elizabeth Marshall Thomas in her 2006 classic *The Old Way: A Story of the First People*, the stories are (or once were) commonly mythlike accounts of actual hunts. They were recited over and over by men in special voices, becoming almost chants, to which everyone listened. There follows one such story, in the actual words of the hunter, of how an antelope is taken down by a poison arrow. I especially like it, even in full translation, because it could be a performance made one hundred thousand years earlier. As paleontologists reconstruct extinct animal species from skeletons, it seems possible to reconstruct ancient social life from such least evolved progenitors.

Ai! What? Is that an ear? Yes, an ear! There's his ear against the sky, he's in bushes, just there, the edge of the bushes. I watch it. Yes, it moves, he turns a little, a little, hi! he lifts his head, he's worried, he sniffs, he knows! He looks, I'm down low, down low, just very quiet, down low, he doesn't see me! He's safe, he thinks. He turns around. I am behind him. I creep forward, eh! I creep I creep, I am

just that far, eh! just that from me to there, quiet, quiet,
I'm quiet, I'm slow, I have my bow, I set the arrow. Ai!
I shoot. Waugh! I hit him! He jumps. Ha ha! He jumps!
He runs. He's gone! I shot him. Right here, just here the
arrow went in. He jumped, he ran that way, going that
way, but I got him.

Storytelling, including especially recorded tales of successful hunts and epic adventures, consumed 6 percent of the overall recorded time during the day, but they consumed 81 percent of the evening. The overall effect was to convey the big picture of the group's existence. It united them into a rule-based community with a single culture. As the elder Di/xao explained in an earlier Ju/'hoansi account, "Our old people long ago had a government, and it was an ember from the fire where we last lived which we used to light the fire at the new place we were going."

3

LANGUAGE

The Ju/'hoansi are fully human. They have a history they carry in their heads. They know who they are. Their massive forebrain is balanced unsafely on the same slender vertical neck as in any city-dwelling citizen. Their species—our species—is uniquely endowed with language, the greatest evolutionary advance since the eukaryotic cell.

A very few animal species have rudimentary culture. A local troop of Japanese macaques have learned, from the example an innovative female in their midst provided, how to clean sweet potatoes by washing them in water. Equally impressive, members of at least one chimpanzee troop use bush stems stripped of leaves to fish for termite soldiers, the suicidally aggressive insect fighters I mentioned earlier that bite and hold on to any invader of their nest. Members of a second group of chimpanzees have learned from one another how to swim and dive or otherwise move through water. These are among the very rare examples of true cultures—behavior invented by individuals and groups and passed on by the social learning of others. But no animal species, at least none out of the more than one million known, has a language. What then is language—what exactly? Linguists define it as the highest form of communication, an endless combination of words translatable into symbols, and (this is the important part) arbitrarily chosen to confer meaning. They are used to label any conceivable entity, process, or one or more attributes that define entity and process.

Each society has one or more languages. At the present time there exist about sixty-five hundred languages, of which two thousand are descending in use and in danger of extinction. A few are spoken by no more than a dozen survivors.

Language is necessary for human existence, but in a way wholly different from the service of our spine, heart, and lungs. It is the basis of society, from the simplest to most complex. By inquiry and knowledge made possible, our minds are able to travel lightning-fast through space and time, and, with increasing scientific precision, visit any place on the planet, and beyond. By any measures of liberation and empowerment, language is not just a creation of humanity, it *is* humanity.

Language in the Ju/'hoansi and Manhattanite alike is

the substance of intelligent thought. It recounts episodes of the past and those imaginable into the future—among which choices made constitute decision and we call free will. The mind assembles experiences and constructs stories from them. It never pauses. It evolves continuously. As old stories fade with time, new ones are laid upon them. At the highest level of creativity, all human beings talk and sing and they tell stories.

If language is universal, is it cultural or instinctive? Many independent studies of child development have demonstrated that language is both. That is to say, language is globally the same as the form in which its capacity began. In the vocabulary acquired, on the other hand, speech is almost completely learned, hence varies drastically from one culture to another. Yet, among even the culturally advanced societies, the use of emotional coloring by melody and rhythm has remained the same. (For an example of the latter, *Let* me explain, please. Let *me* explain, please. Let me *explain*, please. Let me explain, *please*.)

Grammatical rules, on the other hand, are mostly learned. The theory of a universal grammar, famously advanced by Noam Chomsky in the mid-twentieth century, was so complex and jargonized as to escape the indignity of being understood, and has in recent years been largely abandoned for lack of evidence by researchers on linguistic psychology.

The acquisition of language unfolds like any instinct as a predictable series of steps. The key early formative stage of its ontogeny is the babbling of infants. Even newborns as early as twelve hours from the womb react to spoken words but not to other kinds of sounds equally loud. The infantile sounds they subsequently produce are not taught to them but arise autonomously. Even blind and deaf children babble in the absence of external audiovisual stimulation. A few primal words, including *mama* and *papa*, serve as innate attractants for adults, who respond with attention and love.

In adult speech, each of the words may be peculiar to the language spoken, hence cultural in origin, but tone and emotion have remained hardwired and universal during genetic evolution. People are able to listen to texts in unfamiliar languages yet comprehend the mood of the speaker, a conclusion supported by common experience but also by experimental evidence. In one notable case, psychologists were able to tease out the effect with the use of dramatic stage performances. The experiment, as reported by Irenäus Eibl-Eibesfeldt, a pioneer researcher of human instinct, deserves to be quoted in full:

K. Sedláček and Y. Sychro had the sentence *Tož už mám ustlané* ("The bed is already made") from Leos Janáček's *Diary of One Who Vanished* interpreted by 23 different actresses. In this sentence the gypsy Zefka seduces the young village boy Janiček, whereby her courtship is accompanied by sorrow and resignation. Some of the actresses were asked to express a particular emotion or absence of any emotion (joy, sorrow, neutral, matter of fact), while others were given a spontaneous choice but were subsequently asked what sort of feeling they meant to convey while reciting the line. The recordings of their readings were played before 70 listeners of various origins and educational levels.

The response to the recitations were organized into the following categories: (1) simple statement; (2) amorous; (3) joyful; (4) solemn; (5) comical; (6) ironical and angry; (7) sorrowful, resigned; (8) fearful, frightened. If 60% of all responses fell into the same categories, while the others were more or less diffusely scattered in other categories, the example was considered to be distinctly emotionally colored. The subjective evaluation displayed a high degree of conformity of results. Not only the 70 Czech persons studied but also students from Asia, Africa, and Latin America, lacking knowledge of Czech, accurately determined the effective information from the melodic line of the recitations. In order to compare these subjective judgments with objective data, tone, frequency, amplitude, and sound spectrographs were recorded.

There was a segment of my own life that instructed me in the powers of the language instinct. When I was very young I experienced a form of speech deprivation, the overcoming of which merits—what else?—a story. Eventually the experience taught me a great deal about nature, humanity, and who I really am.

I was an only child whose parents divorced in 1937, close to the depth of the Depression. At that time divorce was still viewed as scandalous, and the economic hardship it caused drove our little family to near poverty. Placed under my father's care I grew up as a near-itinerant, moving almost yearly, attending fourteen schools in Washington, DC, and in localities scattered through the deep southern states: among them Biloxi, Mississippi; Atlanta, Georgia; Orlando and Pensacola, Florida; and Brewton, Decatur, Evergreen, and Mobile, Alabama.

I compensated for this peregrine existence by finding the nearest wild or semi-wild environment within walking or cycling distance, then exploring it for insects and reptiles. When I was ten years old, we landed in a residence only several city blocks from Washington's Rock Creek Park, which I decided to explore armed with a butterfly net and a field guide on insects. I had heroes to inspire me. They include the scientists I heard about dwelling like gods on the upper floor of the nearby National Museum of

Natural History, as well as writers in *National Geographic*. In particular, there was William M. Mann, whose 1934 article "Stalking Ants: Savage and Civilized" was eventually transformative. Thereafter, as my father, his new wife, and I wandered like immigrants across the South, I picked up a few close friends my age but, mostly by preference, explored close by the wildlands alone. Approaching graduation from high school, which at that time happened to be in Decatur, Alabama, I assumed that I would go on to college somehow, then find a career as an entomologist— to remain forever thereafter in the outdoors, exploring unknown wildlands, farther and farther away, eventually to what I called the "Big Tropics," the Amazon and Congo forests.

My prospects for such a career, however, were far lower than I realized at the time. My school records were incomplete, my grades were mediocre and sprinkled with absences and failures. The University of Alabama saved me (and made me a fiercely loyal alumnus to this day). By state law at the time, the university's requirements for admission were only two in number: graduate of a high school, and resident of Alabama. I did well enough to go on to a Ph.D. program at the University of Tennessee in Knoxville, and a year later to Harvard University for the completion of the degree and the rest of my career as a faculty member.

By the time I moved on to the Ph.D. program at Harvard, my boyhood career dreams had grown even stronger. I retained my preference for solitary exploration. With scholarship support, I had the means to travel to what I called the true "big tropics"—the most extensive and least disturbed reaches of Earth with the largest faunas and floras. I was able then and into my postdoctoral twenties to conduct field work variously in Mexico, Central America, the Brazilian Amazon, the Australian outback, New Guinea, New Caledonia, and Sri Lanka.

I found that sustained solitude in strange places without human contact, even for just a few hours at a time, might be physically dangerous but also unproductive of ideas and discovery. Most people, it bore on me at the time without understanding the reason, have a compelling need to talk a lot. They carry on for a while every day, if possible frequently, and, for me at least while alone in wildernesses, constantly. So I evolved an alter ego during solitary field work in the strange distant places I visited. This person had no name, nor was it in any manner an independent entity (I wasn't insane). My alter ego was, simply put, just a shift to a separate frame of mind. It appeared with each heightening of awareness of my surroundings, each forced change in priority of action, wherein I spoke to myself literally in words, but silently. Thus, for example, spake

Alter Ego, tracking a sequence as my disjointed experience along the trail unfolded.

Hold on! Stop! Don't pass by that epiphyte (up on a tree trunk) even if it's hard to reach. There might be something hiding in or on it that's really interesting, an ant colony or God knows what else; you need to see it. [Expletive deleted.] *Can't reach it. Move on.* [Later] *Watch what you're doing! Step real carefully! There could be a ravine behind the heavy vegetation right there ahead on the left. Careful, care—ful, take a look. Wait! Look! Look! There's a column of ants. Something really new. Almost hidden in the litter. Maybe army ants but that's not quite right; Leptogenys (ants), maybe? Move in, move in, careful, careful, it might be something new, something really new.*

Talking back and forth to one another, me and my all-but-silent fellow hunter and chattering adviser, I conducted my field research and absorbed the natural history of the environments I visited. My focus stayed on ants, a wise choice for field research. I came to know that in numbers, in biomass, and in global range, they dominate the insects of their size class, and now by the second decade of the twenty-first century, specialists have come to identify 14,000 species worldwide. I'm not a taxonomist, a biologist who specializes on classification, but because of their exuberant abundance in every place I visited, I was able over a lifetime in the field and among already-collected but unstudied museum accessions by others, to discover and give Latinized scientific names to 450 new species.

My purpose as I silently talked my way through the forests and savannas around the world was to find as many kinds of ants as I could, but among them most importantly new and rare species, and then to learn as much as possible about their social lives—where they nested, their population, castes, their food, their communication system. Each species was unique in anatomy and social behavior, often radically so. Each was a source of new scientific knowledge. I learned about them, I virtually lived with a few of them, and I discovered the precise adaptations they had made to the natural wild environment. I was their storyteller, the first the great majority ever had.

Each one of the ant species, scientists and naturalists will tell you, has its own story, different from those of all the others. Its scouts and battalions went forth on forays, while at home, nurses and builders reinforced the nests and drove off invaders. And if the story has been much the same from day to day, it has unrolled with new chapters through the colony cycles that may take a century to complete. The story is not one of culture but of genetic social evolution stretching across millions of years. Putting the big picture of ant social behavior, species upon species, it is possible to reconstruct some of the history of the modern living world they dominate.

It has taken a lifetime of study, and talking, talking, talking, finally to recognize the tribe to which all people belong. It is the Ju/'hoansi.

4

INNOVATION

What exactly is creative literature, by what means is language rendered as art? And how are we to judge it as such? The answer: by its innovation of style and metaphor, by its aesthetic surprise, by the lasting pleasure it gives. Let me begin by example.

A reader knows he is in the presence of greatness when he reads the following opening lines by Vladimir Nabokov:

Lolita, light of my life, fire of my loins. My sin, my soul. Lo-lee-ta: the tip of the tongue taking a trip of three steps down the palate to stop, at three, on the teeth. Lo. Lee. Ta.

To shed light on the infinite importance of literary style, I believe it useful to add for comparison the first lines of Jonathan Franzen's *The Corrections*, winner of the 2011 National Book Award and widely praised for its innovative style:

> The madness of an autumn cold front coming through. You could feel it: something terrible was going to happen. The sun low in the sky, a minor light, a cooling star. Gust after gust of disorder. Trees restless, temperatures fall-ing, the whole northern region of things coming to an end. No children in the yards here. Shadows lengthened on yellowing zoysia. Red oaks and pin oaks and swamp white oaks rained acorns on houses with no mortgage.

While I realize I have no credentials as a critic of fiction, this sounds to me like a painfully ostentatious display of learning by a Harvard sophomore with a great deal of promise. One gets the feeling that as literature this long book may not lift off the runway. For some it does; for me it does not. The Franzen novels thus are characterized by a sweeping simulated exactitude. Their protagonists are steered through a jumble of brand names, undefined technical terms, allusions to philosophy, and whatever else the author has come across in his ruminations and can add to his literary stone soup. They belong to a category the critic James Wood has called hysterical realism. In their fractured stream of consciousness there exists little awareness of or even interest in the depths and roots of human nature.

But now let me add a perspective that in fact finds value in *The Corrections*, in Franzen's subsequent, well-regarded novels *Freedom* and *Purity*, and in similar postmodernist works. They are ethnographic, giving fine-grained descriptions of the personalities and histories of dysfunctional Midwestern families. They are gossip of course (Franzen sounding as though he is your talkative friend), which is why people so love biographies and fiction in the first person. The enjoyment is innate and very Darwinian, having evolved from the aforementioned Paleolithic campfire talk. Postmodern narrations and for that matter all fiction worth its mettle, does what science cannot: it provides an exact snapshot of a segment of culture in a particular place and time. The productions are like photographs that preserve for all time not just the people as they actually seemed, looked, or even truly were, including their dress and posture and facial expressions, but also the surroundings most important to them—their homes, their pets, their transportation, their trails and streets. Who can fail to be riveted by the oldest surviving photograph, taken by Joseph Nicéphore Niépçe in 1826 or 1827, the exact date uncertain, of so mundane a scene as a Burgundy rooftop—immediately to be followed by street scenes and even a pedestrian standing, waiting at a curb next to an empty street—for what, until when? It strikes you then of how far back you have been taken in time. Lincoln and Darwin are still teenagers at this moment, Florida is still a wilderness, and no European knows the source of the Nile.

Fine novels and antique photographs are pixels of history. Put together, they create an image of existence as people actually lived it, day by day, hour by hour, and in the case of literature, the emotions they felt. Finally, they trace some of the seemingly endless consequences that followed. That is why we so value Proust, and why we give John Updike a niche in the pantheon for brilliantly diagnosing, as Updike himself put it, the lives and foibles of an American small-town Protestant middle class. In particular, during the late twentieth century.

Innovation of this kind in the creative arts is important in a second way. The evolution of the arts parallels organic evolution in the manner they both work. The best artists and performers seek original ways to express themselves in image, sound, and story. Originality and style are everything, measured by the degree to which the innovations attract imitation. Thus the challenge of the Salon des Refusés to the Salon de Paris in 1863 is a classic example in the visual arts. Cubism against literalism, marked by Picasso's *Les Demoiselles d'Avignon* in 1907 is another. In popular culture came color motion pictures in 1932 with Walt Disney's *Flowers and Trees*, and Motown in the late 1960s from a blend of soul, blues, and pop. The process is potentially eternal. More and still more: breakthroughs will race the heart of the innovators, stiffen their spine, encourage their search for breakthroughs. By the late twentieth century, experiments in new techniques and styles in all the creative arts were multiplying exponentially. Absurdly wild abstract art was displayed against a background of mystifying atonal music. Originality for its own sake had joined the masters of the creative arts.

For the origin of cubism, Picasso gave metamorphosis as his central goal: "Any artist worth his weight," he said,

> must give to objects to be represented the most possible plasticity. Thus for the representation of an apple: if you draw a circle, it will be the first degree of plasticity. It is possible, however, that the artist wishes for a greater degree of plasticity, and in this case the represented object will end up depicted in the shape of a square or cube which in no way negates the model.

The drive for innovation can be viewed as an analogue of genetic evolution, and to good effect. Cultural evolution adapts our species to the inevitable and constantly changing conditions of the environment. Its innovations are the equivalent of mutations in the genome. These biological accidents have occurred throughout the history of humanity, in the same manner and at the same degree as in other species. Mutations are very diverse. They are individually rare, and in the great majority of cases either harmful (hence the hundreds of unhappily familiar hereditary disorders such as color blindness, cystic fibrosis, and hemophilia) or neutral, having no detectable effect on health or reproduction. In time they disappear or remain at most at a very low frequency, the latter by coexisting at the same site as silent recessives to dominant beneficial genes. Only a tiny fraction of mutations are successful, benefitting the individuals that carry them and spreading through the population as a whole. Sometimes they have enormous consequences. One example is the ensemble of mutant genes that prescribe lactose tolerance. A small random change in DNA base pairs thus made possible milk consumption, and the dairy industry thereafter spread throughout most of the world. Another case is the mutant sickle-cell gene, which in double dose causes lethal anemia but in single dose protects against equally lethal malaria.

The unsuccessful and neutral genes we all carry around in our bodies are called by geneticists the mutation load. From their midst, by further mutation and changes in the environment that perchance favor them, has come the biology of the human organism as it exists today. We can value no less the innovations, only a few of which prove successful, that drive the creative arts.

AESTHETIC SURPRISE

Serious art, whether expressed in score, script, or image, seizes you on first encounter. It then holds and distracts you long enough to lead your mind away and through the remainder of its content—perhaps to understand the whole intended meaning, perhaps to revisit a fragment for sheer pleasure. The overall feel of a creative work (call it the signature) may come at the beginning or at the end, and sometimes only after the experience when it is stored in long-term memory and is the first thought that comes to the conscious mind upon recall.

The signature introduces the aesthetic surprise, whether merely beautiful or otherwise instinctively more deeply compelling. In visual art, for example, it exists in both the parade of the tall ship sails and the apocalyptic tilt of the sinking *Titanic*; Gustav Klimt's surreal gold setting in his portrait of Adele Bloch-Bauer that seems to melt the lady into the metal; Francis Bacon's exaggerations in his brutal self-portraits, which shout the damning consequences of complete honesty; and the lithograph of the great race-horse Citation, 1948's Triple Crown winner, lifted by C. W. Anderson in full triumphant stride, to be compared with the horrific screaming steed in Picasso's *Guernica*.

Creative artists, playing back and forth along the aesthetic spectrum from beauty and majesty to horror and death, use defining signature features to catch and hold your attention. In traditional landscape art, a typical example is Alfred Thompson Bricher's placement of startling white foam atop a breaking wave against a somber dun-and-olive shorescape. In abstract art, Hans Hoffmann's *Magnum Opus* presents a glaring yellow rectangle against a large bright red splash, with a mysterious dark blob added to the side. The eye is forced to travel: yellow to red, then to black. To what purpose is left to your subconscious mind.

The instinct to both present and respond to a simple identifying feature is not unique to humans. It is the equivalent of the "sign stimulus," or "releaser," identified by behavioral scientists as universals in the living world. In an early textbook example, male stickleback fish develop a red belly during the breeding season, which advertises their territory and warns off rival males. They do not need a whole male with a red belly to initiate aggression; just a red spot on a moving object will do. Researchers have used red spots on dummy fish with a wide variety of shapes, including bare eclipses and circles, to evoke attacks. The red spot is the sign stimulus.

The same holds true for the olfactory senses. Male moths are drawn to very specific chemicals released into the air by waiting females of the same species. Hundreds of species of moths may be communicating on the same night, without confusion, because each uses the exact chemical signal (the sex pheromone) of its own kind. When the same substance is applied to formless dummies in tests, males of the correct species emerge from the night, then not only land on the dummy but try to mate with it. Even bacteria exist that aggregate and exchange genes with others, providing they emit the same kind of signal.

Sign stimuli, or at least signals and ensembles of signals that serve the same function, are similarly part of the human psyche. Their presence is confirmed by the existence of another phenomenon discovered by animal behaviorists: the supernormal stimulus. It is well known that when the eggs of herring gulls fall out of the ground nest, or else are removed and placed there by a scientist, one of the parents rolls it back into the nest. What is less well known, even to most naturalists, is that when two artificial eggs are placed side by side outside the nest, the parents attend to the larger one first—even if it is abnormally large. When the experiment is ramped up, the bigger ones continue to win, even if the preferred dummy egg is so large that the adult gulls are forced to climb on top of it.

Human beings are not quite that dumb of course, at least most of the time, but we are ruled much more by instinct than most realize. An hereditary bias, for example, has been revealed in the way people judge facial beauty in young women. For a long time it used to be assumed that the most attractive face is the average in each dimension of a great many faces in a healthy population. However, when this notion was tested with judgments tallied from lifetime residents in North America, Europe, and Asia, it was found to be close but not quite right. The most beautiful face has a slightly smaller chin on average relative to the rest of the face, eyes set farther apart, and higher cheek bones. Model agencies, Hollywood casting firms, and big-eye-anime artists have known this for a long time.

Because innate preferences do not come out of the blue, it is natural that evolutionary biologists would ask *why* this one exists. The search for ultimate cause is called "Darwinian." What advantage in survival and reproduction, if any, we ask, might result from such a facial configuration? One possibility is that the image is a sign of juvenescence: the possessor is more likely to be younger, hence virginal and with a relatively longer-lived reproductive potential.

The same general principle holds in literature. Consider

the aesthetic extremes of emotion by listening first to Emily Dickinson.

> *It was not Death, for I stood up.*
> *And all the Dead, lie down—*

Close to the opposite end of the spectrum is the shout from Walt Whitman's seaman:

> *O Captain! my Captain! our fearful trip is done,*
> *The ship has weather'd every rack, the prize is won . . .*

You've been aroused, you know what is coming, you will remember what Dickinson and Whitman felt when they put pen to paper.

Often a great aesthetic force in one mode of expression can be joined with another mode to magnify the same subject. We find an example in Alexander Gilchrist's description of William Blake's illuminated manuscripts. When Gilchrist discovered and unveiled them in 1863, he set out to do them justice:

> The ever-fluctuating colors, the spectral pygmies rolling, flying, leaping among the letters; the ripe bloom of quiet corners, the living light and bursts of flame . . . make the page seem to move and quiver within the boundaries.

And sometimes a description has a compelling beauty, even when like much of visual art, it exaggerates the factual claims within it. Such is the exquisite closing by F. Scott Fitzgerald of *The Great Gatsby*.

And as the moon rose higher the inessential house began to melt away until gradually I became aware of the old island here that flowered once for Dutch sailors' eyes—a fresh, green breast of the new world. Its vanished trees, the trees that had made way for Gatsby's house, had once pandered in whispers to the last and greatest of all human dreams; for a transitory enchanted moment man must have held his breath in the presence of this continent, compelled into an aesthetic contemplation he neither understood nor desired, face to face for the last time in history with something commensurate to his capacity to wonder.

The understanding of the creative arts by critics, as experts in the subject are called, tends to approach through stages. Their reviews of a particular work by its signature often features a comparison with the artist's earlier work and reputation. It holds the readers' interest until (in longer reviews) the details of its content are scanned. Next may come contemplation of what the artist intended, taking into account his or her life story and the circumstances leading to this particular work. Finally there is judgment, a summing up with placement on a scale ranging from dismissal to fulsome praise. Reviews and criticisms can themselves be works of art, albeit of a separate kind. Brahms' Second Symphony is a great work of art; its analysis by Reinhold Brinkmann is an excellent example of art criticism.

Some of the best signatures within the creative arts do not just surprise the aesthetic sense, they astonish it. The best way to raise this bar is to follow a statement immediately by its contradictions. This device used by Charles Dickens to open *A Tale of Two Cities* may never be surpassed.

> It was the best of times, it was the worst of times, it was the age of wisdom, it was the age of foolishness, it was the epoch of belief, it was the epoch of incredulity, it was the season of Light, it was the season of Darkness, it was the spring of hope, it was the winter of despair, we had everything before us, we had nothing before us, we were all going direct to heaven, we were all going direct the other way—in short, the period was so far like the present period, that some of its noisiest authorities insisted on its being received, for good or for evil, in the superlative degree of comparison only.

In yet another medium, photography, Rachel Sussman's collection *The Oldest Things in the World* presents candidate

trees and other plants thousands of years in age. Like rare
human supercentenarians (110 years old or older), they
tend to the horizontal, sprawling, gnarled, and asym-
metrical, yet they inspire awe, forcing us to think back to
the reality of the vanished time in which they spent their
youth. Seeing these thoroughly remarkable individual
organisms brings with it a troubling negative correlation:
many of the species to which the ancients belong are also
very rare, some close to extinction. The international
champion in both categories is a 43,000-year-old King's
holly (*Lomatia tasmanica*) living in Australia, which if cor-
rectly dated is not only the oldest recorded but also the
last of its kind.

Fables and fairy tales are rich hunting grounds for
such existential clashes. An amazing print by Ben Carlson exhibited at the National Museum of Wildlife Art in
Jackson Hole, Wyoming, shows a lion overcoming and
about to devour a wolf. It illustrates a fable about the folly
of pride:

A wolf left his lair one evening in fine spirits and an
excellent appetite. As he ran, the setting sun cast his
shadow far out on the ground, and it looked as if the wolf
were a hundred times bigger than he really was. "Why,"
exclaimed the wolf proudly, "see how big I am! Fancy
me running away from a puny lion! I'll show him who

is fit to be king, he or I." Just then an immense shadow blotted him out entirely, and the next instant a lion struck him down with a single blow.

To link the humanities through the creative arts to science is a difficult exercise. Why should we even try? The creative arts are among both the most intellectually advanced and most ephemeral of human endeavors. "The arts are true to the way we are and were," Helen Vendler has written in minimizing the prospect of a synthesis, "the way we live and have lived—as singular persons swept by drives and affections."

So far so good, but then Vendler adds, ". . . but not as collective entities or sociological paradigms."

Therewith she conjures the magic in the unknowable, once called by Nietzsche the colors at the edge of the rainbow. She chooses Joseph Conrad, who spoke of "that mysterious, almost miraculous, power of producing striking effects by means of detection, which is the last word of the highest art." She appends the source of her own conviction that we take poetry in particular directly, unclassified, exactly as the poet conceived it: "All my later work has stemmed," she concludes, "from the compulsion to explain the direct power of idiosyncratic style, to convey the import of poetry."

What a lovely journey Helen Vendler has taken, and

how well she has marked her trail for others to follow. Nevertheless, art criticism needs to excavate much more deeply. It stands to make a lot more sense, most certainly amplified with knowledge originating in science. Otherwise the creative arts will continue to grow like trees sprouted outside the forest, less than a part of the living world ecosystem.

The solace of familiarity, cited here as a metaphor for the fail-
ing of both the humanities and the sciences. (*Lamppost*, William
F. Smith, 1938. Metropolitan Museum of Art, New York.)

II

The humanities, particularly the creative arts and philosophy, continue to lose esteem and support relative to the sciences for two primary reasons. First, their leaders have kept stubbornly within the narrow audiovisual bubble we inherited happenstance from our prehuman ancestors. Second, they have paid scant attention to the reasons why (and not just how) our thinking species acquired its distinctive traits. Thus, unaware of most of the world around us, and shorn of their roots, the humanities remain needlessly static.

LIMITATIONS OF THE HUMANITIES

Until a better picture can be drawn of prehistory, and by that means the evolutionary steps that led to present-day human nature can be clarified, the humanities will remain rootless. Human nature, the centerpiece, is not the genes that prescribe it. Nor is it just the traits of culture most widespread in present-day human populations. It is the hereditary propensity to learn certain forms of behavior and to avoid others—what psychologists call "prepared learning" versus "counter-prepared learning." Among many examples of prepared learning that have been thoroughly documented, infants pick up language obsessively, and as older children they are prone to play in ways that imitate adult behavior. We are counterprepared, on the other hand, to trust strangers or enter unknown dark forests, and we form a lifetime dread of snakes and spiders each by as little as a single frightening encounter.

During the biological evolution of our species, the origin of language evidently preceded music, and both language and music evidently preceded visual art. Is that timeline correct, and, if so, what are the implications? How thereby are the emotions stirred by literature, music, and art related to one another? From studies on evolutionary change in other species, we know that intermediate stages, the "links" of evolution, often produce mosaics. That is to say, some traits are advanced, others have reached an intermediate state, and still others have changed scarcely if at all. When, for example, in 1968 I examined the first discovered primitive ant of Mesozoic age, preserved in amber ninety million years old, I found it to have been in the middle of major mosaic evolution. This "missing link" between ancestral wasps and modern ants had wasp mandibles, an ant waist, and antennae intermediate between the ancestral wasp and descendant ants. I gave it the scientific name *Sphecomyrma* ("wasp ant").

To what level then did modern humans evolve these and other capabilities at the time of the breakout from Africa and spread around the world? And why did they so evolve? The full meaning of the humanities will not come from STEM (science, technology, engineering, mathematics). It will come from a combination of many less vaunted disciplines, of which the most important are what I call the Big Five: paleontology, anthropology, psychology, evolutionary biology, and neurobiology. These fields of research are the friendly ground of science, where the humanities will find full and ready alliances. They will encounter some of the same in astrophysics and planetology also, but chiefly as mega-theaters for the play of human emotion, because they have no way to explain its meaning.

The main shortcoming of humanistic scholarship is its extreme anthropocentrism. Nothing, it seems, matters in the creative arts and critical humanistic analyses except as it can be expressed as a perspective of present-day literate cultures. Everything tends to be weighed by its immediate impact on people. Meaning is drawn from that which is valued exclusively in human terms. The most important consequence is that we are left with very little to compare with the rest of life. The deficit shrinks the ground on which we can understand and judge ourselves.

History, conceived in the conventional sense, is a product of cultural evolution. Historians are scholars who tease apart the proximate causes of cultural evolution—in trade, migration, economics, ideology, war, leadership, and fashion. They have successfully taken us back to the dawn of the Neolithic, when agriculture was invented and food surpluses and villages added, thence chiefdoms, paramount chiefdoms, nations, and empires. All this change together, we understand, drove the cultural evolution that created the modern world. But history thus truncated is incomplete without prehistory, and prehistory falls short without biology. The Neolithic revolution began only long enough ago, about ten thousand years, to produce a few changes in small ensembles of genes among the newly settled populations. That amount of time is way too little to explain the hereditary and environmental origin of human nature itself. As populations spread around the world, they carried intact the basic genome prescribing human intelligence and the fundamentals of human social behavior.

Sixty thousand years to ten thousand years before the present—the approximate time span of global settlement by humanity—is the equivalent of, very roughly, five hundred generations. That is still not enough to account for the origin of those traits that hold us together as a single species, including our bizarre hairless and bipedal bodies, our globular skull crammed with an oversized brain, our apish emotions. Then, of great consequence, there was the common instinct to generate languages composed of arbitrary sounds and meanings. Also, there was the shared capacity to practice creative arts, as well as explore and innovate control of the environment, and invent creation myths that reinforce tribal religions.

I think it fortunate that five hundred generations have proved too short a time for our human species to split into multiple species, each reproductively isolated from the other, not hybridizing but with each generation diverging further apart. Such multiplication had been common in our older, prehuman ancestors. The moral and political problems presented would have been insoluble. Only the extermination of all but one could have worked, which is how our species (*Homo sapiens*) evidently dealt with our sister species, the Neanderthals (*Homo neanderthalensis*).

Human beings are not only weak in grasping time but nearly unconscious of what is going on around them in present time. In our daily lives we imagine ourselves to be aware of everything in the immediate environment. In fact, we sense fewer than one thousandth of one percent of the diversity of molecules and energy waves that constantly sweep around and through us. The part perceived is just enough to safeguard our personal survival and reproduction, and for the most part appropriate for the stresses endured by our Paleolithic ancestors. Such is the way that evolution by natural selection works. We are the product of a force that is both powerful and maximally parsimonious.

Biologists use the German word *Umwelt* (roughly, "the world around us") to denote that part of the environment we are able to perceive by our unaided senses. The Umwelt was all our prehuman ancestors needed for millions of years to deal with the African savanna environment, but it was enough. We survived, while other hominin species related to us, but with different perceptions and bad luck, did not. The same is true of condors patrolling the high Andes, with telescopic vision and a keen sense of smell; of hagfish above the abyssal benthos, forever lost in midnight darkness, but geniuses at detecting faint traces of rotting flesh; or agelenid spiders, crouched far back in their funnel-shaped nests, alert to the slightest tug on a silken thread that betrays the passage of an insect prey.

What then is the human Umwelt, and how did it come to be, and why? These are among the central questions of both science and the humanities. A quick answer to the first part is that our species, having evolved as bright children of the African savanna, is adequate in several sensory modalities, weak in most, and a complete blank in others.

We are primarily audiovisual, one of the few animals on the planet, along with birds and a smattering of insects and other invertebrates, that depend on sight and sound to find their way. In vision, however, the only particle to which we respond is the photon. Still more restrictive, our photoreceptors detect no more than razor-thin slices of the electromagnetic spectrum. Our vision begins at the low-frequency end with red (and does not even extend earlier to infrared) and it ends short of ultraviolet at the high-frequency end. If we had a better set of photoreceptors, there would be a potentially vaster spread of colors and shades of colors for us to behold and name. If we could add the vision of hawks and butterflies, the impact on the visual arts would be revolutionary.

And sound? It is essential to our communication, but by comparison with the auditory geniuses of the animal world, we are close to deaf. Bats of many species swoop and spin through the air and pluck fast-flying insects with almost unimaginable precision. Even more impressively, the bats don't rely on sounds from the insects: they make high-frequency sounds of their own and locate the prey by the echoes that bounce back to them. Some kinds of moths are equipped with ears tuned to the bat emission frequency, and they are programmed to drop to the ground the instant they hear the clicking of the bat's echolocation. Other bats detect ripples on the surface of the water and snare fish by raking their claws through the water. Vampire bats of tropical South America smell their way at night to resting mammals—including humans who forget to close the windows—make small slices through the skin, and lap up the blood trickling out. (Perhaps the time has come to write another, more chiropteran *Dracula* epic.) At the opposite end of the sound-frequency spectrum, elephants rumble in complex conversation too low in frequency for our ears.

What about smell? Humans by comparison with the rest

of life are virtually anosmic. Every environment, natural or cultivated, is alive with pheromones, chemicals used to communicate among members of the same species, and allomones, used by organisms to detect other species as potential predators, prey, or symbiotic partners. Every ecosystem is an "odorscape" of still unimagined complexity and brilliance. (Allow me to say "unimagined" for the odor and taste environment, since humans have almost no vocabulary for chemical reception.) When all invertebrates and microorganisms are counted in, ecosystems contain from thousands of species to hundreds of thousands of species. We live within a natural world held together by odors.

Even a trained naturalist walking through forest and meadow has no idea of the thunderous round-the-clock chorus of olfactory signals, its varying mixes forming a riot of airborne scent, beyond your ken and mine but not theirs, not the forest dwellers primed to receive and upon whose perception their lives depend. Below the surface, other pheromones seep up through soil and leaf litter; the whole in time are picked up by gentle breaths of air that brush the ground surface. Like invisible smoke, they disperse and disappear.

Scientists studying the chemosensory world, I among them, have been impressed by how closely the pheromone molecules fit their functions in the species using them. The size of the pheromone molecule, the rate at which it disperses, the time and place it is released, and the sensitivity of other members of the same species ordains how far the signal carries. It also fixes the amount of privacy needed. Consider a female moth calling for a mate. Her sex attractant must be unique to her species. Small quantities must travel far—up to kilometers out in some cases—and it must be read and trigger a response in the right kind of mate, not another kind, or worse, a spider or moth-hunting wasp.

So what do the phantom worlds in which we actually exist mean for the humanities? Surely we cannot picture the living world and keep it safe without understanding the soundscapes and odorscapes by which it is organized.

As a naturalist, I am reminded by the humanlike condition of the pleuston, the collection of organisms adapted to live exclusively in a two-dimensional ecosystem, the surface of water. Riding the surface tension like acrobats on a safety net, they are a strange assemblage of microbes, algae, fungi, and tiny plants and animals.

Only a few relative giants live in this molecular-thin slice of Earth's biosphere. Among the most conspicuous are the water striders, members of the "true bug" order Hemiptera, to which also belong shield bugs, assassin bugs, leafhoppers, scale insects, and aphids. All are distinguished by the possession of sharp-tipped proboscises, which they use to pierce and withdraw fluids from plants and animals. As ferocious predators, water striders rule the pleuston, competing with fish below and dragonflies and birds above for insects and spiders that accidentally fall into the water. They have evolved in precise detail for life within the pleuston: canoe-shaped bodies, three pairs of long, spindly legs specialized in turn, the rear for balance, the middle for speed, and the front directed forward from the head and lined with teeth for seizure of the prey with mantis-like strikes. Their middle and rear legs together stretch far out, distributing their body weight enough that their feet dimple the water tension but never break it. Their entire bodies and all the appendages are densely coated with microscopic water-repellent hairs. Nothing, not the fall of rain (each drop the equivalent of being flushed with a firehose), not the splash of waves, and not even being pushed beneath the surface, can wet their bodies.

Little wonder that a common name for water striders is Jesus bugs. By one criterion these denizens of the pleuston are enormously successful. With ancestors dating back at least one hundred million years, into the Age of Dinosaurs, water striders are today represented by more than two thousand species distributed in different, overlapping ranges over most of Earth's surface. One group of species, members of the genus *Halobates*, are the only insects of any kind known to live either within or upon the open sea.

The global pleuston is exquisitely adapted for the perpetually flat world in which it exists. Its member species never leave except for short trips from one body of water to another. Upward and downward trips into other realms of existence are rare and imperfect. Pleuston dwellers are largely unresponsive in body and instinct to the rest of reality. The skin of water, along with what comes in and what goes out, is the universe they know.

The water striders, princes of this realm, seem so very strange to us, but only as we are to them as perceived by their sense organs out of their world into theirs. Our bodies are specialized for the ecosystem in which our species evolved. Our minds are accordingly limited. Our hope for full self-understanding depends on knowledge not just of ourselves but of the specialization of the other living systems around us.

Is there a place for creative arts in the invisible codes and rhythms of the millions of species that share the planet with us? Perhaps in music and the visual arts? And what possibilities might await in synesthesia, the blending of sensory modalities with one into another, say chemical into auditory or visual? Consider this surmise one step further. We may in the near future be able, with brain-science technology, to read the minds of at least songbirds, apes, and reptiles, next butterflies, ants, and water striders. Then it will be possible to simulate their *Umwelten* with virtual reality.

For the moment, however, we are physically trapped inside the humanities bubble, and worse, remain unconscious of its limitations. The grotesquely lopsided content its restrictions force on us is illustrated vividly by comparing the number of words used in different languages to classify sensory response in different modalities. Start with our celebrated Ju/'hoansi, the Kalahari Desert Bushmen, who are hunter-gatherers with a social organization and daily activity schedule thought to have been characteristic of the distant ancestors of all present-day humanity. In the full known vocabulary of 117 words in combined Ju/'hoansi dialects specifying senses, 25 percent are devoted to vision and 37 percent to hearing, but only 8 percent refer to either smell or taste. This disparity should be of no surprise, since like the rest of us the Ju/'hoansi are relatively weak in smell and taste.

The rest of humanity is remarkably similar in its sensory vocabularies. In the languages of the Teton Dakota Sioux, Zulu, Japanese, and English-speaking people, words applying to vision varies from 25 to 49 percent of the whole, while only 6 to 10 percent apply to smell and taste combined.

The isolation of our specialized, audiovisual species is even more severe when compared with other animals. We are close to "blind" when speaking of touch, humidity, and temperature. Some kinds of freshwater fishes use electric fields both to communicate with one another and to hunt prey. We witness the activity with technology, but otherwise are oblivious to it (unless we grab one and suffer a potentially fatal shock). Nor can we detect Earth's magnetic field, used by some bird species to navigate during their yearly migrations.

The shortcomings of the creative arts and humanistic scholarship that are becoming more apparent in the age of STEM have resulted, even in the outer reaches of science fiction, in an extreme anthropocentrism. Nothing, it seems, counts except on the impact on people. One result is that we are left with little to compare and therefore to understand and judge ourselves.

To summarize, the humanities suffer from the following weaknesses: they are rootless in their explanations of causation and they exist within a bubble of sensory experience. Because of these shortcomings, they are needlessly anthropocentric and therefore weak in their ability to recognize the ultimate causation of the human condition.

Protagoras of Abdera, in the Fifth Century BCE, declared that "Man is the measure of all things." That world view was challenged in his time, and it should be more so today. We need another perception in order to get the story straight. It should read, "All things must be measured in order to understand man."

THE YEARS OF NEGLECT

A little hymn I first heard sung by a tenor in a church in Pensacola, Florida, when I was fourteen years old, brought tears to my eyes and compelled me to ask for baptism and formal membership in the Southern Baptist faith. That, I was told, is the way to affirm belief in Jesus and join Him forever in Paradise.

> *Oh that old rugged cross, so despised by the world,*
> *has a wondrous attraction for me;*
> *for the dear lamb of God left his glory above*
> *to bear it to dark Calvary.*
>
> *So I'll cherish the old rugged cross,*
> *till my troubles at last I'll lay down;*
> *I will cling to the old rugged cross,*
> *and exchange it some day for a crown.**

* *United Methodist Hymnal.* By George Bennard, 1913.

Any child reaching the age of reason can understand the promise: a transformation of the social mind, a masterpiece of proselytization. In just one stanza and the chorus, "The Old Rugged Cross" captures suffering, love, redemption, and community, together the heart of evangelical Christianity.

It also serves as a reminder of how the humanities, of which the study of religion is an integral part, differ fundamentally from science in mode of thought. The humanities alone create social value. Their languages, buoyed by the creative arts, evoke feelings and actions instinctively felt to be correct and true. When knowledge is deep enough and all set in place, the humanities become the preeminent source of moral judgment.

But—wait a minute! Some things are inherently good, others inherently evil, are they not? It may seem that way, but it is also true that every thought and every action must be placed in a context, both scientific and humanistic, before it can be morally judged.

Consider nuclear weapons, a risk to all life and a curse upon the Earth. On the other hand, the dropping of two atomic bombs stopped World War II in the Pacific and, at least as Americans see it, saved millions of American and Japanese lives. Subsequently the fear of nuclear exchange constrained the Cold War and national wars in general. In the interpretation of world history, where does the solution to this moral conundrum lie? How might we go about finding it?

Enter the humanities.

Science owns the warrant to explore everything deemed factual and possible, but the humanities, borne aloft by both fact and fantasy, have the power of everything not only possible but also conceivable.

Because the humanities also consist of everything that is human, all that is human comprises the humanities. It should be axiomatic that education of the young consists of a wisely chosen balance between science and the humanities. Such a curriculum was once called a rounded education; now it is usually called a liberal education. The idea of a liberal education provided all citizens has been one of the greatest achievements of the American democratic tradition.

The idea was well stated shortly after the beginning of the republic by Thomas Jefferson, in his 1818 commission report for the establishment of the University of Virginia. All should receive an education, Jefferson wrote (putting aside the glaring hypocrisy of his slaveholding), both to provide a citizen's own livelihood but also to improve his morals and faculties. "To understand his duties to his neighbors and country," Jefferson continued, "and to discharge with competence the functions confided to him by either; To know his rights . . . And, in general, to observe with intelligence and faithfulness all the social relations under which he shall be placed."

The ideals for public education as expressed by Jefferson have remained the core of the American tradition to this day. Yet the humanities have become the weak sister of the sciences in the esteem and support it receives from the American people.

In 2010, conscious of this disparity, a bipartisan group of members of both the Senate and House of Representatives called on the American Academy of Arts and Sciences to prepare a report on the condition of the humanities and social sciences in the United States, and to evaluate their role in American life and education. The model was the 2007 report sponsored by the U.S. National Academies, *Rising Above the Gathering Storm*, which evaluated the status in America of the STEM disciplines (science, technology, engineering, and mathematics). The focus was to be national, but the conclusions reached could be expected to have global implications.

The central committee of the American Academy was drawn from senior leaders of universities, learned societies, government agencies, and cultural institutions. Their final report, released in 2013 and entitled *The Heart of the Matter*, reached far beyond the Jeffersonian prescription to bring educational philosophy into the modern era.

The Heart of the Matter confirms that for all our frivolities and foibles, we are (how to put it?) a decent people.

We overreach, we boast, we fumble, we have an inordinate fondness for guns. Our most celebrated heroes are not poets or scientists; few Americans can name even a dozen of either living among us. Our heroes instead are billionaires, start-up innovators, nationally ranked entertainers, and champion athletes.

America has come increasingly to identify itself by celebrity and money. Even so, Americans of all socioeconomic groups hold as a first principle that education of high quality should be provided for everyone. As a serious test of that presumed unanimity, what do business leaders think of liberal education, with a balance of science and the humanities? Of course, Shakespeare doesn't sell Toyotas. Yet, surprisingly, a 2013 online survey conducted for the Association of American Colleges and Universities found that three in four business leaders would recommend the concept of a liberal education to their own children or to another child they know personally. All agreed that liberal education is important to some degree. Fifty-one percent ranked it as very important, 42 percent fairly important, and only 6 percent somewhat important.

Moreover, Americans respect the creative arts. At one level of sophistication or another, we depend on them as a perpetual source of both entertainment and—however otherwise we may put it—intellectual enrichment. Much of what we value the most is of high quality and growing better. The fine arts of Bel Canto and symphonic cadenzas are familiar companions to rock, folk, country, and western, even if much less attended. Great visual art is accepted as authentically great, and well worth seeing in the original. A survey made by the National Endowment for the Humanities found that for the period 1982–2008 the percentage of Americans visiting an art gallery or art museum at least once a year ranged between 20 and 25 percent.

In short, the humanities are widely viewed as immensely important for society, and accordingly esteemed. But institutions identified with the humanities do not receive support and operate anywhere close to their value as thus subjectively judged, while universities, like Yale, now stress the sciences in accepting freshmen and creating new courses.

The overwhelming problem is poverty, and a lack of respect. The humanities almost never receive enough funds to finish the projects to which their artists and scholars aspire. They have relatively few long-term wealthy sponsors of the Renaissance mold. Monasteries and other religious retreats no longer serve them as creative sanctuaries. They are classified as luxury items in national and state budgets. They offer too few jobs for the legion of inspired young men and women yearning to devote their lives to some form of the arts and humanities.

Above them both, casting a deep shadow like some alien mother ship parked above Manhattan, are the natural sciences. In the United States, from 2005 through 2011, the physical and biological sciences, along with mathematics, consistently received 70 percent of support for academic research and development from the federal government. The medical sciences and engineering did almost as well, with just above 60 percent of their needs. Educational research together with the behavioral and social sciences came in close to 50 percent. At the bottom were the humanities, which, with the exception of law, stayed close to 20 percent. The rest of support for the humanities came principally from academic institutions.

Science and technology have been supported massively by taxes from the American people for what is generally considered the public good. The knowledge obtained from this largesse is a contributing reason for America's global economic and technoscientific dominance.

The humanities, in contrast, are supported primarily by educational institutions, which receive their income from tuition and endowments, along with a tax-based sliver from government. In the competition between science and the humanities for funds provided by the American people, the humanities rank consistently lower than science.

Americans typically assign prestige of an occupation by the salaries provided its professionals. A good measure of this valuation is the annual starting salary paid college graduates. The U.S. Department of Labor reported in 2014 that the new STEM professionals were kings, receiving between $50,000 and $80,000. Liberal arts graduates, namely in architecture, English, elementary education, journalism, and psychology, were at the bottom, with starting salaries at $40,000 or less.

Americans are often reminded that research and development in basic science are good for the nation. That is obviously true. But it is equally true for the humanities, all across their domain from philosophy and jurisprudence to literature and history. They preserve our values. They turn us into patriots and not just cooperating citizens. They make clear why we abide by law built upon moral precepts and do not depend on inspired leadership by autocratic rulers. They remind us that in ancient times science itself was a dependent child of the humanities. It was called "natural philosophy."

Why then are the humanities kept on starvation rations? Partly because so much of our available resources are appropriated by organized religions. The vast majority of people around the world belong to one particular religious faith or another, which is defined not so much by a belief in God as by its idiosyncratic creation myth. Each member is committed to believe that his religion's creation myth, accounting for the supernatural origin of the universe and humanity, is superior to all others. The problem is that all the myths cannot be correct; no two can be correct; and almost certainly none is correct.

Across the centuries organized religions have created transcendent music, literature, and art. The most moving ritual I have witnessed (if I may be indulged) is the Roman Catholic Easter celebration of Lumen Christi. It begins with a darkened cathedral filled with the faithful. The rear door opens, and the bishop enters holding a lighted candle. He calls to the still unseen assembled, "Lumen Christi"—the light of Christ. He then walks slowly down the center aisle, followed by attending priests. The faithful, standing silently, each holds an unlit candle. The episcopal group lights their candles row by row, until the cathedral is fully illuminated. At the altar, the Easter service begins.

Gripped by such majestic displays evolved over centuries, it is easy to forget that religious art is dedicated to the regnant creation story, that deviation from the story is not permitted, and that brutal wars have been fought to replace one story over another. The secular humanities, to put the matter plainly, must compete with organized religions and religion-like ideologies for attention and volunteer public funding. The first is free to explore and innovate; the other is not.

Competition from faith-based culture is not the only force suppressing the humanities. Far more potent is the digital revolution. Science and technology are not hostile to the humanities. They bend them to no supernatural dogma or blind ideology. Nevertheless, their competition has become overwhelming. Artisanal manufacture, garden agriculture, and commercial wild fishing are shrinking toward extinction. Automation, mass production, and global communication increase the world economy, not the influence of the humanities, such that the best jobs are taken by those trained in science, technology, and technology-aided commerce and law.

STEM has become the American symbol of power, the equivalent of Rome's SPQR (Senatus Populusque Romanus). There seems almost nothing a technoscientific culture cannot accomplish in time—cure every disease, create artificial body parts and organisms, produce unlimited food in vertical LED illuminated hydroponic farms, desalinate seawater with solar or fusion energy. Leaders in brain science and artificial intelligence have begun to search for the origins of mind and spirit, once territories exclusive to the humanities.

It is self-evident that to succeed in the new technoscientific world, people need high-quality education—and lots of it. Very few countries have been able to meet the challenge the new reality presents. In particular, the United States, having led the world in the STEM juggernaut, has slipped to a low point in teaching it to the young. According to a 2013 study published by the Organization for Economic Cooperation and Development, adults aged sixteen to sixty-five in the United States ranked only twenty-first out of twenty-three top-level countries in mathematics and seventeenth out of nineteen countries in problem solving. Does it matter? Might innovation and development still be accomplished by a small but highly educated elite? Not likely, attendees of a Capitol Hill conference were told in 2013. Even then, an estimated 2.5 million STEM jobs were not being filled because of the shortage of well-educated workers. Already a large majority of even low-level jobs required at least elementary computer skills.

The problem before America is being magnified by a growing inequality of income, correlated with a sharp decline of the middle class. Americans in all socioeconomic classes now recognize that just to stay afloat, and evolve, and thereby to flourish in what is rapidly becoming the new American dream, it is well to know the full lay of the land ahead. I will next argue that this expansion requires not just the promotion of STEM but also an equally powerful new growth within the humanities.

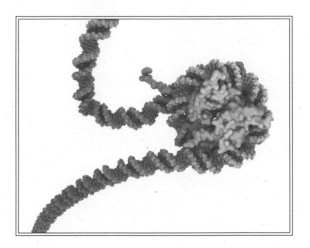

The humanities and science address segments in the single continuum of creative thought, from the molecular processes of heredity to the emotional responses they program, reaching through all space and time. *Above*, the underside of an African swallowtail (*Papilio lormieri*), produced by the evolution of one particular genetic code. (Photograph by Robert Clark.) *Below*, a segment of DNA with a fragment of the histone proteins around which DNA is wound. (Image created with Molecular Maya [Clarafi.com] by Gaël McGill, precisely as it would appear if viewed by ultra-magnification.)

III

Science and the humanities share the same origin and brain processes of creativity. They can be drawn closer together and widely joined in substance through a more thorough application of five disciplines—paleontology, anthropology, psychology, evolutionary biology, and neurobiology—bound together by the evolutionary process in heredity and culture.

8

ULTIMATE CAUSES

The best art criticism is often brilliant, rising from deep intuitive wisdom. Nevertheless, due to the extreme subjectivity of the subject, the insights easily slide across the surface and off target. We are told by Wallace Stevens, for example, that Picasso meant for cubism to serve as a metaphor of the fractured disorder of the modern world. Picasso himself explained that cubism was to possess "flexibility," with the figure on canvas depicting a stage of development capable of transformation to other stages.

The intent of the artist is intensely idiosyncratic and difficult to assess. For his part, Matisse allowed that he was in a fury to create new perceptions, delivering repetitive shocks of the new. Gauguin assembled his Tahitian masterpiece, *Where Do We Come From? What Are We? Where Are We Going?* as a panoply of the human life cycle. He then set out to commit suicide in the hills above Papeete, but changed his mind and moved to the Marquesas Islands. In the Rothko Chapel at Rice University, the artist's last work, subjectless and formless, represents, according to the master critic Robert Hughes,

> an astonishing degree of self-banishment. All the world has been drained out of them, leaving only a void . . . the viewer is meant to confront the paintings in much the same way as the fictional viewers, gazing on the sea in a Caspar David Friedrich, were seen confronting nature: art, in a convulsion of pessimistic inwardness, is meant to replace the world.

In the hothouse climate of experimental arts and criticism, it is not surprising that bizarre subcultures sprout abruptly and randomly, like courageous mushrooms and dandelions on an otherwise well-tended lawn. They defy coherent explanation itself: Dadaism, hyper-realistic cans of tomato soup, postmodernist philosophy and literature, heavy metal and atonal music. Whether emanating from ordered or disordered minds, they give us glimpses, still disordered unfortunately, of the emotional checkpoints and decision centers of the subconscious mind.

It is time for a deeper probe in a different setting, entered at a different angle, to a greater depth, and exploring a deeper causation. Why have the creative arts so dominated the human mind, everywhere and throughout history? We will not find the answer in the finest art galleries and symphony halls. The innovations of jazz and rock, arising more directly from human experience, will probably give us a better idea of where to excavate. Because the creative arts entail a universal, genetic trait, the answer to the question lies in evolutionary biology. Bear in mind that *Homo sapiens* has been around about 100,000 years but literate culture has existed for less than a tenth of that time. So the mystery of why there are universal creative arts comes down to the question of what human beings were doing during the first nine-tenths of their existence.

Preliterate societies of pure hunter-gatherers and primitively agricultural hunter-gatherers still surviving have a great deal to tell us about the natal millennia of prehistorical culture. Their lives may look simple. They don't watch television (most don't, yet!), or search the Internet, or drive to the supermarket for groceries. Still, the Kalahari Ju/'hoansi, to take my favorite and one of the hunter-gather societies closely studied, know the geography of their territories like a street map, and back and forth across hundreds of square kilometers—and within them every fruit tree, waterhole, potential campsite, and lookout hillock. Their vocabulary may be very small compared to that of modern urbanites, but they can name and describe plants and animals with expertise rivaling that of taxonomy-trained naturalists. Their conversations and storytelling, focused on business in daylight and on most everything else in firelight, are diverse and detailed, and in the case of the Ju/'hoansi sprinkled with words abetted by three kinds of clicks, each made in a different part of the air passage.

So what can we learn from the Ju/'hoansi and more generally, the Old Way of life? As background, each human being, with no exception, belongs to a single, reproductively isolated species of animal confined by its idiosyncratic biology and social behavior. Before this characterization can be read as a hint of divine interpretation, be aware that the same can be said of hundreds of other social species, from siphonophore jellyfish and web-spinning communal spiders to fellow mammals like porpoises and wolves. Whales grow to great size by seining tiny crustaceans, bats fly at night by echolocation, birds fly at night by the polar magnetic field. Humans think.

From scientists we've learned about the early stages of all this evolution. It turns out that three preconditions came together to produce a human-grade species. The first was the creation of the campsite, made possible by the shift in diet as early as the ancestral *Homo erectus*. A review I have made of the origins of all known complex societies throughout the history of the animal kingdom, representing twenty independent lines in total, revealed that each was preceded by the instinctive construction of nests in which the young were raised with the aid of parental care. For the social bees, wasps, and ants, the nests were variously subterranean or arboreal, and contained special compartments for the developing young. For social thrips and aphids the nurseries were cavities already present within living plants. Social marine shrimp used chambers excavated in living sponges. The nests of early humans were campsites warmed and lit by controlled fire. Thus a widespread but uncommon adaptation, nest building to accommodate progressive rearing of offspring, was the base from which the rare human level could be approached.

This most advanced social trait of social organization, possessed by the twenty evolutionary lines, is "eusocial" behavior, in which division of labor is based not on cooperation among equals but on organized cooperation in which the group members assume long-term roles. To qualify as eusocial in scientific classification, the roles are marked by the programmed superior survival and reproductive success of some of the members. Put simply, altruism exists. A portion of the group members make sacrifices for the good of the group as a whole.

The second precondition for the origin of human societies was high levels of cooperation among members of the group. Each person knew all the others, and something about their labor roles, their ability, and their character.

Division of labor, altruism, and cooperation evolving together put a high premium on social intelligence. In particular, their combination enriched communication. Because the earliest humans were audiovisual, they were able to evolve a capacity for spoken language. The words created were originally arbitrary in meaning and became universal in usage within groups. Sounds are made and fade swiftly. Unlike visual signals they pass through opaque obstacles, if thin, and turn corners. Further, unlike odors and visual signals, words can grow swiftly in numbers, and maximize information transmission. The instinctual animal sounds of our ancestors thereby evolved into human speech. The vocabularies came to differ among groups, but the capacity and driving impetus to talk remained genetically programmed.

Within groups, those with better capacity for language had enjoyed higher survival and reproductive rates than their groupmate rivals. More importantly, in competition between groups, those who won were superior not only in mortal territorial aggression but also in the ability to form alliances, develop trade, and extract materials and energy from sources in the natural environment.

Thereafter the brain grew from 900 cc in the ancestral *Homo erectus* to 1,300 cc or even higher in the early *Homo sapiens*. Viewed from a distance, it should be no surprise that we spend most of our time chattering while scribbling pictures and symbols. And why, in societies too large to know all its members, we treasure and gossip about celebrities.

BEDROCK

Humanity is a chimaeric species. The physiological basis of our senses and emotions remains much the same as in our apish ancestors. Our capacity for the creative arts—narrative language, dance, song, pictures—dates before the African breakout over sixty thousand years ago. Everything else has changed. Science and technology are doubling, according to discipline, every ten to twenty years. They address everything in the universe, all space and time and everything on Earth and on every conceivable star system and exoplanet. The humanities, on the other hand, are stuck with people.

Therein lies the dilemma of the humanities. They, and with them a large part of liberal education, are expected to describe and explain the essence of a social world of Paleolithic emotions, mediaeval institutions, and god-like technology with no clear idea of meaning or purpose. In the search for meaning, science and technology have roles disjunct from the humanities. Science (with technology) tells us whatever is needed in order to go wherever we choose, and the humanities tell us where to go with whatever is produced by science. Where science has created new modes of intellect and vast material power, the humanities have addressed the issues of aesthetics and value or lack thereof mostly raised by the relentless Stakhanovite energies of science.

The human enterprise has been to dominate Earth and everything on it, while remaining constrained by a swarm of competing nations, organized religions, and other selfish collectivities, most of whom are blind to the common good of the species and planet. The humanities alone can correct this imperfection. Being focused on aesthetics and value, they have the power to swerve the moral trajectory into a new mode of reasoning, one that embraces scientific and technological knowledge.

To fulfill this role, the humanities will need to blend with science, because the new mode above all depends on a self-understanding of our species, which cannot be acquired without objective scientific research. Like the sunlight and the firelight that guided our birth, we need a unified humanities and science to construct a full and honest picture of what we truly are and what we can become. That combination is the potential bedrock of the human intellect.

It seems clear, as I've argued, that the humanities can be broadened enough to make the connection in three ways. First, escape the bubble in which the unaided human sensory world remains unnecessarily trapped. Second, sink roots by connecting the deep history of genetic evolution to the history of cultural evolution. And third, diminish the extreme anthropocentrism that hobbles the bulk of humanistic endeavors.

It is common for writers in search of the ultimate meaning of human existence to turn to the mysteries of astrophysics and quantum mechanics. Alternatively, they look to the future map of brain neurons and circuitry. In more traditional modes, many writers seek spiritual enlightenment, thereby to God or to some *mysterium tremendum*, the core of which guides us but stays forever beyond our grasp.

All of these efforts are guaranteed to continue, and fail. They play in one of the most powerful of archetypes, the quest for the ultimate unknown. That quest has taken form through the ages variously in the holy grail, the sword in the stone, the secret code of the ancients, messages left by extraterrestrials, the key to the inner sanctum, or the physics-based Theory of Everything.

In truth, self-understanding of the deep, genetic qualities of our species has already been attained, a substantial part of it anyway, with the long path to the rest laid out in principle. Many surprises await, especially in molecular and developmental genetics, but few that add to the paradigm shift needed to explain who we are and what we can expect to attain without leaving planet Earth. The answers thus far, as I've stressed, are being provided by the relevant bedrock disciplines we would logically expect. They comprise paleontology, anthropology (including archeology), psychology (primarily cognitive and social), evolutionary biology, and neurobiology.

The common thread that unifies this Big Five most relevant to humanity is evolution by natural selection. The preeminence of this universal process was nicely expressed in 1973 by the great geneticist Theodosius Dobzhansky and often quoted since: "Nothing in biology makes sense except in the light of evolution." That claim should now be boldly expanded: Nothing in science and the humanities makes sense except in the light of evolution. As the philosopher Daniel Dennett has put it, evolution by natural selection is the acid that burns through every myth about ordained purposes and meanings.

Biological evolution is defined as the hereditary change of traits within populations from one generation to the next, eventually causing one species to give rise to another, and sometimes splitting into two or more species. Most educated people know this characterization, but very few can tell you much about how it actually works. This spectacular shortfall in education is due mostly to poor science education in schools, and is probably the single most formidable barrier to any effort to bring science and the humanities together to create an authentic general education. The whole is actually quite simple compared, say, to general relativity or to a randomly selected school of abstract modern art. It can be summarized as follows.

The first principal is that genetic evolution (often called organic evolution) cannot be made comprehensible as a single lineage, parent to offspring, thence offspring to grandoffspring, and ever onward backward and forward, ad seriatim. Rather, it is a hereditary change throughout an entire population. Its magnitude is measured as a shift in the percentages of competing traits within the population as a whole. The population comprises freely interbreeding individuals. It may be an entire species or a geographically isolated part of a species. It may occupy an offshore island, for example, where other populations of the same species live on the mainland.

Among the genes of individuals making up the population, mutations are constantly occurring, albeit at a very low rate for any particular gene. One in a million per generation in a population is not an unusual figure. A mutation broadly defined is a random change that occurs by an addition or subtraction of the DNA letters making up the genes, or by changes in the numbers of the genes, or even by a shift in the locations of the genes on the chromosomes. Mutations can affect any biological or psychological trait, large or small.

If the change in the traits prescribed by a mutation prove relatively favorable in the surrounding environment to the survival and reproduction of the individuals carrying it, then the mutant multiplies and spreads through the population in competition with other genes located at the same site on the chromosome. An aforementioned human example of one such mutation prescribes lactose tolerance, which renders milk and all its products palatable and makes possible the cultural invention of dairying. If, on the other hand, the traits prove unfavorable in the surrounding environment (say preserving lactose intolerance), the mutant gene will either persist in a very low percentage of individuals in the population or disappear entirely. Traits with significant impact on human health persisting in marginal numbers include thousands of rare genetic diseases and disorders such as Duchenne muscular dystrophy, haemophilia, and a propensity to develop certain cancers.

Evolution is a change in the population in the frequencies of competing genes affecting the same traits. When, for example, the frequency of lactose-tolerant genes increases by even just several percent, evolution has occurred. The competition is between those newly arrived as mutations versus those already in place. Because winning or losing among alternative genes is almost entirely determined by the environment, biological evolution is said to occur by "natural selection," the phrase introduced by Charles Darwin to distinguish the process from artificial selection, which is the breeding of plant and animal strains by human choice.

Evolution by natural selection occurs continuously in every population of every species, either changing the frequency of genes or holding them steady. At one extreme, its pace is fast enough to create a new species in a single generation—as, for example, by a simple doubling of all the chromosomes and all the genes they carry. At the opposite end, evolution is so slow that some traits of the species have remained similar to those of ancestors living tens of millions or even hundreds of millions of years ago. With only a little exaggeration, these laggard species are often called "living fossils." Familiar examples with traits two hundred million years old or older include horseshoe crabs, dragonflies, beetles, and cycads.

In the elementary theory of evolution by natural selection, the unit of heredity that mutates is the gene, and the target of natural selection by the environment is the trait prescribed by the gene. The distinction might be made more memorable with the aid of Hollywood science fiction horror movies: the monstrous giant mutants of insects and crocodiles that escape from secret government laboratories cause local mayhem, but they have no chance to reproduce and spread in Darwinian competition with their normal progenitors. Or so we hope.

There has been unnecessary confusion, especially by popular writers on evolution, between individual-level selection and group-level selection. The problem arises from an erroneous distinction between the unit of heredity and the target of selection. The problem is simply resolved when referred to the well-developed discipline of population genetics, as follows: Individual-level selection acts on the traits that affect a *group member's* survival and reproduction, apart from its interaction with other members of its group. Individual selection predominates in the early stages of social evolution, when many of the hereditary traits affect the success of the individual independently of its interactions with its groupmates. The individual may, for example, live alone during part of its life cycle. Among its *groupmates*, it may take larger shares of food and space for itself and for its offspring.

Group-level selection affects the traits that are interactive with groupmates, so that the success of an individual's genes depends at least partially on the success of the society to which the individual belongs. In the most advanced social organizations, characterized by a sterile worker caste and illustrated by siphonophores, which resemble jellyfish, as well as ants, bees, wasps, and termites, group selection almost entirely overrides individual selection.

So, where does humanity fit in on the spectrum of individual-to-group selection? We are located near the center. As a consequence, human nature is driven by conflict between individual selection, which promotes selfishness on the part of individuals and their immediate families, and group selection, which promotes altruism and cooperation put in service to the larger society. This admixture in levels of selection, forged in an arena of instinct and reason, is part of what makes humanity unique.

The role of group selection in social evolution is consistent with the proven fundamentals of population genetics. Its occurrence throughout the animal kingdom is supported by massive evidence from the field and laboratory. It has nevertheless been challenged by an alternative explanation of social evolution called *inclusive fitness theory*. In essence, the theory says that social behavior evolves in accordance with the degrees of kinship among members of the group. The closer their kinship, the more likely the group members will share their resources and cooperate in labor. The price for each of the group members in survival and reproduction is compensated by the increase of genes identical to its own shared with others in the group by kinship. If you sacrifice your life for your extended family, it matters more than if they are only distantly related or not related at all. More of your bravery genes will persist in the population if you make the sacrifice for close relatives.

At first thought, this concept of kin selection, extended beyond nepotism to cooperation and altruism within an entire group, appears to have considerable merit. I said so when I first synthesized the discipline of sociobiology in the 1960s and early 1970s. Yet it is deeply flawed. In spite of the excited attention at first given to it, no one has ever succeeded in measuring "inclusive fitness," as its core property is called. To succeed, not only would it be necessary to determine pairwise kinship throughout the group, but also to assess reciprocal gains and losses in fitness through time. Beside the technical difficulty, the equations offered to conduct the overall analysis have proven mathematically incorrect. There is a basic reason for this mistake. In the purview of inclusive fitness theory, the individual group member, not the gene, was made the unit of selection. Inclusive fitness means how well the individual does in its relation to every groupmate, discounted by the percentage of the shared appropriate genes, through its entire reproductive life. However outwardly attractive, there exists no evidence of such a process, or any need for it to explain the origin of advanced social behavior.

I admit I've come down strongly on this issue, knowing full well that a proper scientist is supposed to express some degree of uncertainty and speak in terms of probability. But there is a need to cut through the confusion caused by the dwindling corps who support the inclusive fitness theory, and place less emphasis on the theoretically grounded and documented process of group selection.

Why does this disagreement matter? It is impossible to overestimate the importance of group selection to both science and the humanities, and further, to the foundation of moral and political reasoning. This is a subject we need to get clear and straight.

As Darwin first observed in *The Descent of Man* (forgive me for an appeal to authority), competition among groups of humans has been a major contributor to traits universally considered noble, that is, those manifesting generosity, bravery, self-sacrificing patriotism, justice, and wise leadership. To explain these qualities with individual selection alone, it is necessary to take a thoroughly cynical view of humanity, based on selfish genes and the convoluted methods of deception and manipulation they prescribe. Common sense tells us there is something ruling humanity that is a great deal better. We admire the band of brothers in battle, the fireman who risks all, the anonymous philanthropist, the embattled teacher in Appalachia. Heroes are real and all around us. Their good deeds are the safety net of civilization.

Because of group selection, and its obvious consequences in the evolution of human social behavior, there is reason to suppose that the better angels of our nature need not be drilled into us under the threat of divine retribution, but are instead biologically inherited. We are by a fortuitous consequence of the fundamental principles of natural selection far more than trained savages.

It is equally the case that the love of nature, leading to the floras and faunas within the wildlands, comes easily to people everywhere, even those whose entire lives have been spent in cities. It would be a mistake to justify parks and reserves primarily as economic assets, or even as outdoor health clinics. Wildlife conservation stands on a moral foundation independent and sufficient unto itself.

The origin of morals or lack thereof is suggested by the fable of the scorpion and the frog, ancient in origin. A scorpion wishes to cross a stream, but cannot swim, and asks the frog for a ride. The frog demurs, saying the scorpion might sting and kill it. The scorpion argues that would not happen, because both of them would be killed. So they start off, and at midstream the scorpion stings the frog. As they both sink, the frog asks the scorpion how it could do such a terrible thing. The scorpion replies, "Because it is my nature."

BREAKTHROUGH

It is becoming increasingly clear that natural selection has programmed every bit of human biology—every toe, hair, and nipple, every molecular configuration in every cell, every neuron circuit in the brain, and within all that, every trait that makes us human.

Natural selection as grand master of evolution means that humanity was not planned by any super-intelligence, nor was it guided by any destiny beyond the consequences of our own actions. The human product has been tested and often revised in every one of the thousands of generations of its geological lifespan. Success for our evolving species meant survival during each reproductive cycle in turn. Failure would have resulted in decline on the road to extinction, thus ending the evolutionary game. It has happened in the vast majority of other species, many before our very eyes. The last passenger pigeon and Tasmanian marsupial tiger died in cages, the last great auk was killed by egg poachers, and the last ivorybill was seen being chased by a crow across the canopy of a remote Cuban forest.

The same fate could have befallen our ancestors any time in the past six million years. Like every other species surviving today, ours was just extraordinarily lucky. Over 98 percent of the species that ever lived have vanished, to be replaced by the multiplying daughter species of the survivors. The result has been an approximate balance, extinction versus birth, in the number of species evolving from one epoch into the next. The history of any particular lineage is a journey through a constantly changing labyrinth. One wrong turn, one misstep of evolution, even a single unfortunate delay in an evolutionary adaptation, could have been fatal. The average life-span of a mammal species during the Cenozoic Era, across the time in which our prehuman ancestors lived, was about half a million years. The lineage that eventually led to modern humanity split from the common chimpanzee-human ancestor about seven million years ago. Its luck held thereafter. When prehuman populations declined during hard times to perhaps a few thousand individuals, and many of the species related to us fell to zero, our lineage wound its way across the six million years of the Quaternary Period. It continued as a perpetually evolving entity. Occasionally it split into two or more species. All continued to evolve, but only one persisted as the line which—perchance—led to *Homo sapiens*. The other sister species continued to evolve, diverging from the prehuman line. In time, each died or split into daughter species of their own. Eventually, all declined and disappeared.

Through most of the six million years following the chimpanzee-human split, as many as three species, probably more, classified together as the australopithecines, coexisted in the African homeland. They were basically vegetarian, perhaps consuming a small amount of meat when opportunity provided, as do modern chimpanzees (with about roughly a three percent caloric intake). The species evidently varied in the kind of vegetation they consumed. Those relying on coarser, more fibrous plants evolved heavier jaws and teeth. The specializations taken together represent what evolutionary biologists call an adaptive radiation.

Out of the radiation, one lineage shifted to heavier consumption of meat, especially that cooked by lightning-struck grassland and savanna ground fires. In the early stages, the groups invented campsites, at first no more complex than a bird's nest. To these they added controlled fire—no more difficult than carrying the ember of a burning tree limb from one place to another.

Out of this elementary but eventually momentous shift

came *Homo erectus*, the direct ancestor of *Homo sapiens*, no later than two million years before the present. That ancestral species lasted until at least 100,000 years ago. By then, the brain of at least one of its populations had become much larger and the jaw and dentition smaller and lighter.

The final transition to *Homo sapiens* was well under way during the geologic life-span of *Homo erectus*, but more likely it originated not in that species but earlier, in its own apparent direct ancestor, *Homo habilis*. The fossil evidence recovered for the habilines is much sparser than for *Homo erectus*, and also for later transition species immediately antecedent to *Homo sapiens*.

It was evidently in *Homo habilis*, present in Africa between 2.3 and 1.5 million years ago, that the swerve began that ended in modern humanity. In its segment of prehistory, the cranial capacity, hence brain size, rose from 500 to 800 cc, well above the size of the modern chimpanzee. It grew again, to that of *Homo erectus* (about 1,000 cc) and thence to *Homo sapiens* (averaging 1,300 cc or more). The momentous threshold was crossed by the early *Homo sapiens*: the larger brain provided larger memory, leading to the construction of internal storytelling, then for the first time in the history of life to true language. From language came our unprecedented creativity and culture.

We are still evolving, not by direct selection leading to

larger brains and higher intelligence, but by homogeniza-
tion of what we have developed through interbreeding
around the world. The average genetic diversity between
populations is decreasing steeply while the total genetic
diversity of humanity is staying about the same. In biology, as in culture, we are becoming an evermore united
species.

GENETIC CULTURE

The exponential growth of brain size launched during the habiline period of prehistory around two million years ago was the most rapid transformation in the complexity of an organism in the history of life. It was driven by a unique mode of evolution, called gene–culture coevolution, in which cultural innovation increased the rate at which genes favoring intelligence and cooperativeness were spread more rapidly; and, acting in reciprocity, the resulting genetic change increased the likelihood of cultural innovation.

By a widespread consensus, the scenario drawn by scientists thus far begins with the shift by one of the African australopiths away from a vegetarian diet to one rich in cooked meat. The event was not a casual change as in choosing from a menu, nor was it a mere re-wiring of the palate. Rather the change was a full hereditary makeover in anatomy, physiology, and behavior. The body was slimmed, the jaws and teeth shrunk and grew lighter, the skull swelled and assumed a globular shape. The society changed, from a band wandering in the manner of chimpanzees through a protected territory, with its members foraging for food either independently or in small troops, into a larger, better coordinated assembly comprising teams of hunters and gatherers. All left and returned, in the manner of wolf packs, to fixed dens and rendezvous. These habilines, possessing free hands and greater intelligence, likely learned to bring fire to the site, then keep it alive and controlled.

This theoretical reconstruction has gained traction from fossil remains and the lifestyles of contemporary hunter-gatherers. Meat from larger prey was shared, as it is by wolves, African wild dogs, and lions. Given, in addition, the relatively high degree of intelligence possessed by large ground-dwelling primates in general, the stage was then set in prehuman evolution for an unprecedented degree of cooperation and division of labor. These traits were also favored by heightened competition in social skills among group members, leading to natural selection at the individual level, and increasingly enhanced by competition between groups. Natural selection at the group level especially favored altruism and cooperation.

The result that logically follows from these processes is a positive reinforcement between cultural evolution and genetic evolution. Each, alone, could be expected to increase the rate of brain growth. Together, the two would experience mutual positive reinforcement. There followed exponential growth of the brain, from the habilines to the neanderthaloids and modern humans—slow at first, then faster and faster until a limit was reached due to the counterforce created by physical limits on the relative size of the cranium. A simple anatomical circumstance finally lowered and halted the advance of human genius. The whole human organism was not designed, especially during the long era of stone knapping and spear throwing, to support an infinitely more massive bobble head on an increasingly slender neck beyond that attained by the primitive races of *Homo sapiens*. It stalled about three hundred thousand years ago with the best that could be done.

The occurrence of gene-culture coevolution is fundamental to the unity of science and the humanities. Consider, as an example, the process of aging. Why must we die? More generally, why does each species in turn, ours included and each hereditary strain within each species, of which humanity has plenty, possess a characteristic lifespan? If you seek a long-lived dog as a companion, as opposed say to a sheep herder or wild-boar hunter, you would be better off with a Chihuahua (lifespan twenty years) than with a Great Dane (lifespan six years). Plants also have programmed lifespans. Some northern conifers live on average about a century, magnolias a century-and-a-half, and sequoias and pines of the Southwestern United States for as long as several millennia. But they, too, age and die eventually. What could be more important to both science and the humanities than the human life cycle and preordained human life-spans?

The prevailing theory in evolutionary biology to explain programmed aging and death recognizes that each species of plant and animal has evolved a lifestyle in which most individuals die from external causes—disease, accident, congenital defect, malnutrition, murder, war—well short of their maximum potential longevity. Such was severely the rule in Paleolithic times, when few people reached an age of fifty years. As a result of this idiosyncratic pre-longevity death rate, which is improved but still applies to most humans today, natural selection has front-loaded vigor and reproductive drive. It has programmed the vital physiology and mental state of youth in the youngest adults, while disfavoring older adults. It has bet, so to speak, against investment in middle and old age.

With the dawn of Neolithic civilization, and advent of agriculture and food storage, along with an easing of external mortality, conditions have changed in a way that redirects natural selection along the human life cycle. Thanks to the weakening of Paleolithic mortality factors through cultural evolution, average longevity is increasing and the age of reproduction expanding up to the age of menopause.

One inevitable result in future generations may be an overall population-level hereditary shift and not just an extension of youthfulness and fertility into middle age. The onset of menopause will be pushed back. The impact on both cultural evolution and heredity will increase accordingly.

Gene-culture coevolution probably played a critical role throughout human prehistory. The coevolution follows a constant repetitive cycle. Language and technology edged forward, with the Darwinian gains favoring hereditary lines best able to use them.

In fact, it is possible, I believe even likely, that the habiline revolution—the giant step toward the human condition—was fueled by gene-culture coevolution. During this period of the earliest *Homo*, culture would have created very little technological innovation. If it did, the Neolithic revolution would have come much earlier than was the case, and no stone-age hunter-gatherer societies would have survived to the present time. On the other hand, logic and the accumulated evidence point to gene-culture coevolution as a powerful force in the origin of early oral productive language, along with complex social behavior featuring ever higher levels of empathy and cooperation.

HUMAN NATURE

The human condition—what we are as a species, what we wish to become, and what we fancy we can become in flesh and dreams—depends upon phenomena at four levels. The first is the processing of sensory input, as in hearing, sight, and smell; the second is the reflexes, as typified by eyeblinks and the autonomic nervous system; and the third is paralinguistic, comprising facial expressions, hand movements, and laughter. The fourth and final level is symbolic language, the one capacity that distinguishes *Homo sapiens* absolutely from other creatures. Each of the four levels is altered to some extent by emotional centers of the brain. Subject to decision at the checkpoint centers of the subconscious brain, they summon memories that help form future scenarios in the conscious mind. The result of all these processes is what we call "thinking."

Sensory perception is heavily tilted in humans toward sight and sound, whereas the vast majority of other kinds of organisms rely on chemical cues. As researchers in gestalt psychology long ago discovered, incoming sensory information tends to be distorted and ambiguous in predictable ways. The visual anomalies include the Rubin vase, an image shaped to switch (in our minds) from a vase to a pair of opposed facial profiles, then back and forth between vase and faces. When a Necker cube, composed of six faces of equal area, is turned and tilted, the front and rear vertical edges appear to exchange places fast and repeatedly. The most delusional and frustrating of all is the Müller-Lyer illusion, where, like magic, a V-shaped figure opening the ends of a line makes the line seem longer than an identical line with the V placed closing the ends. The mind strains to verify the physical proof it is programmed to deny.

There are countless other ways in which the visual system copes with the ambiguities of the real world. The brain manages the confusion of visual input by automatically reordering and simplifying the information it provides. When laboratory volunteers were asked to make drawings of figures they had memorized earlier, they proceeded to create shapes more generalized than those experienced. In particular, they improved the symmetry, simplified the figure, enhanced the subdivisions, straightened oblique lines, and isolated incompatible details.

Reflexes, forming the second level of behavior, are truly hardwired and independent of conscious thought, except in later memory. The sneeze is a reflex, as are the knee jerk, involuntary eyeblink, blushing, yawn, and salivation. The most complex reflex is the startle response. Imagine coming up unseen and very close behind someone, almost touching, and making a loud noise (I don't recommend that you actually try this). The person will instantly slump forward, drop the head, close the eyes, and open the mouth. The function of the startle reflex is defense. Think of a Paleolithic hunter who, when stalked from behind and attacked by a predator (say a leopard), instantly rolls forward and away in an overall relaxed posture. Thus correct decision and swift action is attained with no conscious thought needed.

Facial expressions, postures, and bodily movements, the paralinguistic signals composing the third level of human nature, are both consciously displayed and shared to some degree by all cultures. They are also used universally as substitutes or enhancements for verbal language. During his classic field research in the 1960s, the German anthropologist Irenäus Eibl-Eibesfeldt demonstrated in minute detail that people in all societies, from primitive and preliterate to modern and urbanized, use the same wide range of paralinguistic signals. These entail mostly facial expressions, denoting variously fear, pleasure, surprise, horror, and disgust. Eibl-Eibesfeldt lived with his subjects and further, to avoid self-conscious behavior, filmed them in their daily lives with a right-angle lens, by which the subject is made to think that the camera is pointed elsewhere. His general conclusion was that paralinguistic signals are hereditary traits shared by the whole of humanity.

The universal or near-universal fixed-action patterns include the eyebrow flash to express pleasant surprise at an encounter: eyes wide open, eyebrows lifted and mouth smiling. In children and some women, shyness upon encounter is expressed by cutting visual contact, often burying the face with open hands while turning the head away. A playface is assumed by adults with young family children and often includes mock-biting. Play differs as a genetic trait between boys and between girls, with boys prone to engage in imitation fighting, singly or between coalitions, as a way of expressing dominance.

Newborns grasp with hands and feet, and crawl on all fours in search of a nipple. Infants use five sounds: contact, displeasure, sleep sounds (the latter, when given, signals to the mother that all is okay; when absent, something is not okay). Further, a drinking sound also signifies all is well as the infant nurses, and its absence again suggests something is wrong. Finally, crying strongly signifies hunger, pain, discomfort, or fear.

That all of these signals and movements are hardwired genetically is supported by their appearance in blind, deaf, and mute infants. Although these infants have no previous visual or auditory experience, they engage in the same appropriate smiling and crying, plus calmness to denote a neutral mental state. Further, when left alone a while (no tactile contact), they engage in nail biting and facial expressions of despair. If disturbed, they display open palms, a warding-off gesture.

As analyses of the primal form of communication have advanced, so has the observed richness of its universal vocabulary. One suite of signals contains postures that display dominance within groups, as well as the means to achieve it in the first place. They turn out to be similar to those in social Old World monkeys and apes. Deborah H. Gruenfeld, a social psychologist at Stanford University, found that people feel more powerful—and often are so in actuality—when they display the following traits in the presence of groupmates: Behave expansively, keep your hands away from your body, make eye contact as you talk but feel free to look away at your leisure. Don't explain yourself in any detail. Take ownership of the space around you, whether a boardroom or an office cubicle, in order to say to yourself and imply to others, "This is my table, this is my room, you are my audience." Those who practice dominance behavior have higher levels of testosterone and lower levels of cortisol stress hormone.

Another very primitive, even hereditary, signature of dominance is to rest casually in a physical position above your subordinates—whether upon a stage, throne, tower, victory stand, or penthouse. To look down physically upon others, especially in a relaxed posture, is to subordinate them. Not long ago in Madrid, I had two full days to spend in the Prado. I paid scant attention to a special exhibit of Gauguin, being pulled by my plebeian genes instead to gaze wonderingly at the portraits of the Hapsburgs. They are there in all their imperial splendor, each supremely dominant as it was thought by grace of God. One is Spain's Philip IV, depicted posthumously by Rubens, mounted high upon a horse at a rise in the land, attended by an angel in flight above his head. Far away and below are soldiers in a battle of the Thirty Years' War. Little people fighting and dying, Philip removed and calm. He has turned his position partly toward you and gazes impassively downward in your direction.

As a student of innate behavior in animals and humans, I observed with admiration the use of just this technique in one particular colleague over decades at Harvard University. Notoriously weak in research after his arrival at Harvard, negligent in departmental duties and undergraduate teaching, he sustained prestige by an expert performance of dominant status at faculty meetings. Arriving, he would stroll over and sit next to the department head or college dean, chat sotto voce with him, while inspecting his arriving colleagues with steady inquisitorial gaze. Usually unprepared by homework on the subject to be discussed, he would turn possible embarrassment into a posture of command by firmly addressing the chairperson as the meeting began, looking about as though speaking on behalf of the others, "Now, what I want to know is . . ." I kept thinking at the time about this successful strategy, which I had also observed in dominant chimpanzee males. The postures, facial expressions, the chosen lines of sight were closely similar.

Still, social psychologists have learned that the hereditary nonverbal signals can vary a lot in fine detail according to the context in which they are expressed. In 2015, an international team of psychologists led by Paula Marie Niedenthal and Magdalena Rychlowska of the University of Wisconsin discovered through big data analyses that communication by smiling depends in good part upon how diverse had been the people who founded the country in which the studied person dwelt. Smiles that signal friendly intent as opposed to aggressive or competitive intent were more common in countries of diverse origins.

During interactions with strangers, the presence of a smile reliably predicts trust and sharing resources. Moreover, observing smiles that accompany cooperative behavior increases one's cooperation in the future. Negotiating status is another matter. This type of social interaction is complex and potentially disruptive in homogeneous cultures, such as Japan and China, where long-term population stability created conditions favorable to the development of fixed hierarchies. In similar circumstances, a smile can signal that the interaction will not disturb the social order, where specific features of the smile convey derision and criticism amidst other signs of superior status.

Returning to biology, the variability in form and meaning of hardwired signals leads to a part of the fundamental process of gene-culture coevolution: the narrowing and hardening of variation itself as an inborn trait by evolution through natural selection. First called the Baldwin effect, after the American psychologist James Mark Baldwin, who conceived it in 1896, gene-culture coevolution has substantial relevance to the evolutionary origin of human nature. In essence, it is the principle that when a variant in a learned behavior proves advantageous and is repeated often, mutations prescribing rather than leaving it as merely a learned option will increase in frequency, and in time the new trait will become fixed.

The Baldwin effect is strikingly illustrated in the castes of ants and termites. Charles Darwin was fascinated by these insects. He spent long hours seated in his garden at Downs, watching and thinking about the anthills there— so much so a housemaid is reported to have remarked, while referring to William Makepeace Thackeray, the prolific novelist who lived nearby, "It is a pity that Mr. Darwin doesn't have something to pass his time, like Mr. Thackeray."

In fact, the great naturalist was more troubled than Thackeray could ever have been while struggling over any of his plots and dénouements. Darwin had noticed something about ants that appeared to contradict the theory of evolution by natural selection and could prove fatal to it. He knew that typical ant colonies consist of the mother queen plus a few males that if present are temporary guests, and, finally, a great many female workers who run the colony and perform all the labor. His dilemma was this: workers do not reproduce; most are sterile, and in some species they lack ovaries altogether. If worker ants cannot reproduce, and if they are thereby unable to pass on their servile anatomy and behavior, how can these traits have evolved through natural selection? The answer Darwin conceived has proved correct (the man was annoyingly almost always correct). All the female members of the colony, he reasoned, have the same heredity. Which caste each female becomes, whether queen or worker with its distinctive body form and behavior, depends on the environment in which she is raised. Especially important is the amount and quality of food she receives while growing up as a little grublike larva into a six-legged, antennae-waving adult. The details of caste determination vary greatly among the 14,000 known species of ants, but of those studied thus far, the development has proved to be a genetically based version of linear programming. One code solved by researchers has the following rule: If a larva reaches a certain size by a predetermined age, its development continues unabated to the next decision point, producing in the end a new queen, with a full set of wings and ovaries. If instead the larva fails to reach the critical point in time, growth of the tissues destined to become wings and ovaries is shut down, and the adult emerges as a worker—wingless, sterile, and smaller than its sister virgin queens—even when all in the colony brood are genetically identical.

The human equivalent of caste determination, with variation generated by a genetic form of dynamic programming, exists in what psychologists call prepared learning. The phenomenon lies at the base of human instinct and what we all perceive as human nature, and which, in turn and expressed creatively, is the governing core of the humanities.

A textbook example of prepared learning is the fear of snakes—not ordinary fear and caution as one might feel from the approach of a snarling dog or close lightning flashes, but gut-wrenching, paralyzing revulsion. A child can learn to like snakes. They can carry them around fearlessly as pets, as I did earning the nickname "Snake Wilson" at the T. R. Miller High School in Brewton, Alabama, a lovely small town (Clayville in my novel *Anthill*), with a winning football team and only one resident herpetologist, me. I enjoyed showing others that when snakes are habituated to human touch, they will rest in your hands or hunt harmlessly through your clothing, presumably for mice and frogs. Their skin is leathery to the touch, not slimy as many suppose, and their flickering tongue is an innocent organ of smell, not a venomous dart.

But if a child is frightened by a snake—just once and just a little bit and in almost any conceivable manner, even with frightening pictures and stories or the sight of some cylindrical object writhing on the ground—he or she is prone to acquire in one stroke a deep aversion to all snakes, or worse, a full-blown autonomic lifetime phobia with paralyzing effects.

In other words, ophidiophobia looks like an instinct, and in one sense that is the case (as in the traditional sense of the word). But the aversion to snakes is also learned, and easily so, in a manner programmed to be fast and sharply targeted. What was the ultimate cause of the snake-avoiding instinct? The clear answer is the lethal danger human beings and their prehuman ancestors have faced for millions of years. Only a small percentage of snake species are venomous, but those few are the most dangerous animals on Earth's land surface. The highest rate of death from snakebite is in southeastern Asia (vipers, adders, cobras, and kraits). One of the riskiest occupations in the world may be harvesting tea leaves where abundant hidden Russell's vipers lurk—big, aggressive, deadly, and hard to see. Venomous snakes are nearly universal in tropical and temperate parts of the world. Lethal bites, though very rare, still occur even in Finland and Switzerland.

Like ghosts from our geological past, other potentially dangerous animals have imposed their own genetic phobia. Instinctive fear of spiders, arachnophobia, which I have in mild form and never have been able to erase completely, has a sensitive period that begins around the age of three and a half years and persists throughout childhood. Fear of insects generally ("bugs" to the fear-prone) is greater among children at six to eight years than it is later. Prepared fear of larger animals does not ordinarily develop before the age of five, but for dogs (hence in ancient times, wolves) it can appear by the age of two.

It is a remarkable fact that the hair-trigger capacity to acquire conditioned aversions and phobias is almost entirely limited to risks acquired in the wild by our distant human and prehuman ancestors over countless millennia. In addition to various animal enemies, these include close spaces, heights, running water, and exposure to strangers outside home. Our species has not yet had time to evolve phobias to knives, guns, and automobiles, which are by far the most frequent agents of death in modern life.

Drawing closer to the more aesthetic core of the creative arts, measurements of brain alpha waves have revealed that maximum arousal of abstract designs is achieved by approximately 20 percent of redundancy in the elements, roughly the same amount of complexity found in a simple maze, or two turns in a logarithmic spiral, or an asymmetric cross. Less complexity gives the sense of unappealing simplicity, while more complexity appears "crowded." It seems relevant that about the same degree of complexity is presented by a great deal of successful art in friezes, grillwork, colophons, logographs, and flag designs.

The same level of complexity characterizes part of what is considered attractive in primitive art and modern art and design. The optimum complexity principle may be an expression of a limitation in the brain to seize the whole in a simple glance. The same principle is obeyed by the number seven, the quantity of objects that can be counted in one single glimpse—without breaking the image into units, to be counted and added together.

The humanities have yet to come to grips with the chimaeric nature of our minds and creativity. We are ruled by emotions inscribed in our DNA by prehistoric events little known and only partly understood. Meanwhile, infinitely puzzled, we have been catapulted into a technoscientific age that may in time serve instructions to robots well but not the ancient values and feelings that keep us indelibly human.

The endless beauty of life on earth. Each "snowflake" has been composed by the artist from multiple images of species of invertebrates. (The Harvard University Invertebrate Zoology Department members who contributed snowflakes were: Tauana Cunha, Sarah Kariko, Vanessa Knutson, Laura Leibensperger, and Kate Sheridan. Kate Sheridan did the design work of photographing and placing the snowflakes together in Photoshop.)

IV

Even as our species destroys the natural world at an accelerating rate, nature remains a source of deep love and fear. As we hasten to blanket Earth with a humanized environment, we should—we must—pause to consider how and why our relation to nature exists. That degree of self-understanding, as I will next suggest, can be achieved only by a blending of science and the humanities.

13

WHY NATURE IS MOTHER

For almost all of the 100,000 years that humanity has existed, nature was our home. In our hearts, in our deepest fears and desires, we are still adapted to it. Ten thousand years after the invention of farms, villages, and empires, our spirits still dwell in the ecological motherland of the natural world.

We do not and cannot live for long outside of this self-sustaining environment. We exist in a narrow biological niche ultimately dependent on its largesse. The natural world has humbling power and eternal life, so that with good reason we call it Mother Nature. She has been damaged by our activity, but that is nothing new. For the long geologic haul, humanity is just another perturbation. Let her speak for the ages, metaphorically, as did the actress Julia Roberts performing on behalf of Conservation International:*

> I don't really need people but people need me
> Yes, your future depends on me
> When I thrive, you thrive
> When I falter, you falter or worse
> But I've been here for eons
> I have fed species greater than you, and
> I have starved species greater than you
> My oceans, my soil, my flowing streams, my forests,
> they all can take you or leave you
> How you choose to live each day, whether you regard or
> disregard me doesn't really matter to me
> One way or the other your actions will determine your fate
>   not mine.

We are Earth's unruly children who left home to make it big in the city. But as scientists are finding out, and I've stressed, there is a lot of Mother Nature still in our genes. Evolution during our long tenure in the natural world has left an indelible stamp—Darwin's metaphor—in the way we communicate by posture and facial expression.

* "Nature Is Speaking—Julia Roberts Is Mother Nature," Conservation International (CI) (https://www.youtube.com/watch?v=WmVLcj-XKnM).

We are also innately fixed in the choice of the environment we most want to inhabit. Thanks to pioneering research by Gordon H. Orians at the University of Washington and other scientists and humanities scholars, we have a good idea of what may be called the instinct of habitat selection. People tested across diverse cultures have expressed preference for the following home location: placed on elevated land looking out across a broad expanse of savanna sprinkled with small trees and copses with a barrier rise of rocky surface or dense forest to the rear; and finally, next to a lake, river, or other body of water. Their landscape of desire closely approaches the African environment in which our human and prehuman ancestors originated.

Artists from Asia, Europe, and North America show the same combination in their paintings of pastoral and woodland environments. By and large, they avoid primeval environments of the north temperate zone, characterized by dense deciduous and coniferous forests. When such an environment is depicted, it is typically softened by an insertion of meadows and lakes. It leaves space for the instinctive preferences of humanity's ancestors.

Orians has refined the "savanna hypothesis" by including organisms as well. The shape of trees widely used by gardeners—from the temples of Kyoto to the baronial estates of England—possess traits in common with the dominant acacias of the African savannas: they tend to have exceptionally broad canopies relative to their modest height, short trunks, and small, divided leaves. The Japanese have been breeding maples and oaks over a thousand years, aiming to perfect just these qualities. I can testify that, even prior to learning of the savanna hypothesis, my favorite trees were (and they remain) the Japanese maples.

What might be the adaptive advantage of the human preference for a habitat like the African savanna? It is logical to search for one. All known animal species that are mobile are required to seek the environment best suited to their survival and reproduction. They must find their way to the exact suitable place, and get there fast and with unerring accuracy. Why should we not expect to find at least a vestige of this propensity in modern human beings?

I once addressed a convention of landscape architects, during which I emphasized biophilia, the innate love of contact with other living organisms. Biophilia architecture was just then coming into professional acceptance. I included the savanna hypothesis of human habitat preference. I was puzzled by what seemed to me to be their relatively tepid response. Had I been too abstruse or otherwise failed to make my case? Later I asked an architect friend if I had been unclear, and he responded, "Oh no, it's just that we already knew all that stuff."

There is a simple, easily tested reason why the savanna homeland shaped our primal yearnings. From a high

perch, early humans had a wide view of grazing animals and approaching enemies. Living close to a body of
water ensured a supply even during severe droughts. It also
provided an extra source of food. The distinctive physi-
ognomy of the acacias, with low, horizontally spreading
branches, allowed a quick climb to escape lions and other
predators big and fierce enough to take down a human.
The spacious horizontal branches were supports on which
to rest and wait out the pursuers. They also served as a
station from which to scan the local terrain for prey of
their own.

It is within the domain of evolutionary biology to
inquire why most people enjoy a walk in the woods, and
why the experience contributes to sound physical and
mental health. There's the value of exercise, of course, but
there is something else at work deep in our psyche. In our
hearts, we are still in some fashion or other hunters and
gatherers. So follow me back in imagined time to hunt
alongside one of our Paleolithic ancestors.

*In order to survive we must keep our eyes open and our
ears keened. Walk a quarter-mile, cross-country, not along a
trail—predators and enemy scouts like to wait hidden along
trails. A good habitat would be a mature hardwood forest,
anywhere in temperate Europe, Asia, or North America.
You will see plants and animals previously unnoticed; and*

they will be only a tiny fraction of the species alive and unseen a few feet distance to the side: scores of plants, moss, and lichen species; countless fungi, and thousands of spiders, millipedes, centipedes, mites, springtails, and so on through a roster of creatures whose biology is still mostly or entirely unknown. (A handful may contain, even if you are in Washington, DC's Rock Creek Park or New York's Central Park, species waiting to be discovered by science.)

This is the point I want to make: the experience of nature, to those who have learned to absorb it, is a magic well. The more you draw, the more there is to draw. In early visits to wild environments, you will miss most of what is interesting. You will subsequently spot more, and then more, and begin to put names on what you see. Then will come still more new details. Each species, you will learn, is a story in itself. I had an entomologist friend who upon retirement in Florida tracked each Red Admiral butterfly in his backyard generation after generation, caterpillar to adult. This may seem an empty pursuit, but I can tell you that he found Red Admirals to be complex territorial butterflies with individual personalities. His observations had scientific merit.

You need not travel to the Amazon or Congo to experience a wilderness. There exists enough novelty and challenge in what I like to call microwildernesses. A new genre of nature writing has emerged to explore the potential of small plots and fine detail. Annie Dillard was a pioneer in her *Pilgrim at Tinker Creek*, which won a Pulitzer Prize in 1974. Among other notable examples are David M. Carroll's *Following the Water: A Hydromancer's Notebook* (2009); David George Haskell's *The Forest Unseen* (2012); and Dave Goulson's *A Buzz in the Meadow: The Natural History of a French Farm* (2015). These works, combining scientific natural history with poetic interpretation, make the invisible appear, the small grow large, and the beauty of life more plainly experienced in all its dimensions.

Seemingly infinite biodiversity packed into a small space, an image Darwin famously labeled the "tangled bank" in the closing paragraph of *The Origin of Species*, cannot have escaped the attention of at least some creative artists. I have been drawn to Jackson Pollock's well-known 1950 piece *Autumn Rhythm*, which in astigmatic viewing could represent the first glimpse of an overgrown English roadside or other ecosystem, and I was pleased to hear later that this was indeed meant by the artist to be a representation of nature. Whatever else Pollock had in mind for others in his body of work, I had the same sensation, idiosyncratic on my part perhaps, for especially Pollock's *Number 8*.

Creative artists in all fields have nevertheless fallen far short of exploring the potential of wildlands and the bio-

logical diversity within them. There is the evolution leading to the multiplication of species, as well as initial chaos
crystallizing into orderly ecosystems; and invasive species
obliterating ancient natural ecosystems; and many more
natural and human-caused processes that can be trans-
lated into ideas and emotions through evocative poetic
and visual description.

14

THE HUNTER'S TRANCE

Absorption in the tangled bank is the trance an expert hunter enters when searching for a particular kind of plant or animal. Hunting for a single prey, the hunter works like a tiger, alone, rather than in a pack, like a wolf. He is intimate with the fine details of the place and environment in which his quarry lives. He is alert to every subtle change in the ground and vegetation that can betray the route of the animal he seeks. He must be prepared to stalk his target from a distance, placing each foot carefully, as a literal matter of life or death. Or he may wait quietly for hours in ambush. To accomplish a kill, he must know what behavior to expect from the prey—when it can sense him, from what direction, and when it is ready to bolt. The expert hunter performs best when he perceives the *Umwelt* like his prey. For a few, the experience is spiritual, lifting the hunter to a higher level of consciousness.

In *The Hunter's Trance* (2007), Carl von Essen cites a hunter who one day stalked a herd of wapiti elk in the Colorado uplands:

I came across fresh tracks of several elk, including one bull. The morning was bright and sunny, with a slight wind in my face, assuring me of getting a reasonably close approach to my quarry. After perhaps an hour of slow and careful tracking, I came upon a long glade fifty yards wide. If the elk were nearby they would detect my crossing the snowy and slushy meadow. It remained for me to be completely still and pay complete attention to the opposite hillside. I felt now their presence and somehow knew that they felt mine. As I stood there, the sense of time remarkably changed. What seemed like minutes I found later to have been over an hour. An intense feeling of the clarity of the scene swept over me. All my senses seemed to sharpen to an exquisite razor's edge. I heard the tiniest sounds of distant streams and rustling leaves as if magnified in a celestial amplifier. Everything seemed closer to me and I felt, amazingly, a sort of merger of myself with everything, a sense of belongingness. I was connected with everything in that panorama, the grass, trees, rocks, insects, birds, the elk that I knew were quietly moving uphill, out of my sights. I felt a great rush of emotion, a joy at being alive, the chance to exist along with everything else. I will never forget that day.

One does not require a Winchester rifle or lust for a blood ritual over the fallen prey to experience the hunter's trance. As a naturalist I've come close. But even more important to the naturalist than a Zen-like state of total awareness is the *search image*, the combination of traits of a species of plant or animal that allows the tracker to find it as one element among thousands hidden in the vastness of an ecosystem. I cannot find words to describe the pleasure I feel, even to this day, as a naturalist hunting rare and little-known species. But I can tell you a couple of stories.

One of my own best experiences evincing the power of the naturalist's search engine was the discovery of zorapterans, an exceptionally rare and elusive group of insects. One early spring day, during my freshman year at the University of Alabama, I was working my way through the mixed conifer-hardwood forest tract along nearby Hurricane Creek in search of new and unusual kinds of ants. I pulled away the bark from a decayed pinewood log. Beneath were characteristic debris-lined spaces excavated earlier by beetle larvae. This particular niche is a favorite nesting site for small, rarely seen ants and other cryptic inhabitants. A few were there in the log I chose. They began to move away, toward surviving fractions of bark, and dark shelter. Together they formed a mélange, a sparse mix of little scorpion-like schizomid arachnids, springtails, oribatid mites, and undersized anonymous beetles. But something else in this miniature cave especially caught my eye: several little insects vaguely resembling termites, similarly white and elongated, but smaller in size, more delicate, with faster and more erratic movements as they, too, edged back toward shelter.

I collected several live specimens and took them back to my laboratory space in Josiah Nott Hall to examine under a microscope. They were, I quickly learned, a species of the genus *Zorotypus*, an insect so distinctive in anatomy as to deserve placement in an entire insect order of its own, called the Zoraptera, equal in rank within zoological classification to all the flies (Order Diptera), as well as all the beetles (Order Coleoptera), and all the moths and butterflies (Order Lepidoptera).

They are also, I learned, among the rarest of insects, the first having been discovered only in 1918. Since that time zorapterans have picked up the common name "angel insects." They fully deserve the appellation. They are like little lambs racing about, white and pure (the Greek *zor* means "pure"), harmless, and vulnerable to the sharp-jawed predators around them, feeding on fungus spores as we would eat freshly picked mushrooms.

It occurred to me then that one reason zorapterans

had remained hidden for so long might be that they are specialized to live only in the particular microenvironment in which I had by good luck found them. That guess proved correct. I soon was easily turning up my little angel insects as I tore apart the pine logs chosen to be at exactly the right degree of decomposition. A rotting log proved to be the zorapteran search image everywhere I looked. I soon published one of my first scientific articles, "The Zoraptera of Alabama." The report was read, because soon other entomologists were having a kind of sport finding them across much of the eastern American landscape. Might I find zorapterans elsewhere? Yes! Later, engaged in postdoctoral research at Harvard, I casually located the first known zorapterans from New Caledonia and New Guinea in the South Pacific. Others located species in Central America. Later, one of my graduate students, Jae Choe, wrote his doctoral thesis on the life cycle of zorapterans in Panama.

Sometimes extraordinary species are found by careful search, other times they are spotted accidentally in a fleeting glimpse. In 1955, I was in a biologically little-explored part of New Guinea, on the move with a party of local Papuan men, all expert hunters. We were climbing through trackless forest at the base of the Huon Peninsula toward the center of the Saruwaged Range, which rises to about twelve thousand feet. We made the top after five days, and found ourselves in a cold, rainy grassland dotted with palm-like cycads. At the time, I thought I was the first non-native to explore this part of the Saruwaged Range, but discovered later that an American botanist, Mary Strong Clemens, a very tough middle-aged lady, had preceded me in the 1930s. (Surely she deserves a niche in the pantheon of pioneer feminists.)

My aim, apart from the thrill of seeing a mountaintop ecosystem visited by very few other scientists, was to collect ants for Harvard's global research collection. I looked out for other specimens of possible interest (and I did pick up a new species of frog). The ants I discovered changed in species as we climbed through the higher vegetative zones. They also grew increasingly scarce until I could find none at all above 7,500 feet.

Among the few ants I collected as we entered the lower reaches of the mountain forest was a specimen with a bizarre anatomy, walking slowly across a leaf on an understory shrub. I searched all around on the ground and through the surrounding vegetation, but was unable to locate the nest of its colony or find even another individual.

Now, sixty years later, I am ready, with admitted bias, to nominate the ant as one of the most beautiful animals in the world. You may dismiss this judgment because of the specimen's small size, but I ask you first to imagine increasing its five-milligram weight to five kilograms, the latter equal to a large bird or medium-sized mammal. Few of these animals anywhere in the world, and I have seen many in my life, can compete with that little mountain ant.

The chitinous armor of the ant's body is a glistening blackish brown, with a feel to it of colorized polished metal. Parallel rows of furrows run from the forward edge of the eyes to the insertion of the jaws, seeming to form opposing lines of needle-sharp teeth. The darkness of the body gives way to varying shades of shining ferruginous on the antennae and six legs. The ant is female, as are all worker ants, and she must also have been a ferocious warrior. What enemies she faced and what prey she hunted I do not know, and would very much like to learn, because her armament is awe-inspiring. One striking feature is a pair of immense spines that curve back like the horns of a ram, rising from the rear of the midbody, clearly designed to shield the slender waist of the ant. A pair of shorter spines, resembling those of a rose bush, guard the equally vulnerable neck. Two more spines project to the rear from the first segment of the waist across its joint and approach the second segment.

The first specimen of this species was described in 1915 by a German entomologist, who gave it the scientific name *Lordomyrma rupicapra*. The first term includes the Greek expression for "ant." The second, *rupicapra*, borrows the name of the European Chamois "rock goat" (*Rupicapra rupicapra*), because when viewed from above, the lines and shapes of the midbody together are reminiscent of a Chamois head.

The history of the discovery more than a century ago, was but one very small event in an expedition conducted in the grand old style of European natural history, when even less was known of New Guinea's fauna and flora. Robert W. Taylor, a present-day expert studying the ants of the region, with the one specimen captured the record and equally important the spirit of the enterprise. The account is among the best from exploratory natural history, and worth quoting in full.

The *L. rupicapra* holotype was very likely collected at a high-elevation, probably cloud-forested site. The collector S. G. Bürgers was medical officer and zoologist with the remarkable German Kaiserin-Augusta-Fluss Expedition (1912–1913), which for 19 months explored the basin of the Sepik River and its headwaters. The scientific party travelled well upstream of the 560 miles of steamboat-navigable water. Four upriver overland trips averaging 3 months duration were mounted. All Sepik tributaries extending westwards beyond the former Kaiser Wilhelms–Land/Netherlands New Guinea border (that of modern-day Papua New Guinea and Indonesian West Papua) were explored to their headwaters, and the Bewani and Torricelli Ranges (north) and central cordillera (south) penetrated (Sauer, 1915; Behrmann, 1917, 1922). Geographer Behrmann's narratives mention specimen collecting on a "damp, cold 2,000 meter high peak," and crossing over "the 2,000 m Schrader Range" (ca 4°59′ S, 144° 05′ E). Nationaal Herbarium Nederland collection-locality records of the expedition botanist C. L. Ledermann include four sites above 1,000 m elevation on Mt Lordberg (4°50′ S, 142°29′ E), 1,000 m; Mt Hunstein (4°29′ S, 142° 42′ E), 1,350 m (where 17 days were spent at the summit); and "Holrungberg, 1,800–2,000 m." The cited coordinates are those of expedition camp sites.

I might have left the discovery of this species unremarked here, except the account of its capture combines two archetypes resident in both science and the humanities. The lure of geographical exploration and scientific discovery that brought me to New Guinea can still be summoned in Earth's remaining wildlands. They cannot help but call to the creative arts, pleading for closer attention to the living natural world.

To the same end, I'll close my account of the hunter's ethos with a story of my meeting in 1952 with Vladimir Nabokov, who was my natural history soulmate. I was a twenty-three-year-old graduate student, working on my Ph.D. thesis in Harvard's Museum of Comparative Zoology. Nabokov was fifty-three years old, his fame as an English-language novelist yet to come. I knew of him only as a lepidopterist, a specialist on butterflies. His expertise was on the blues—small, beautifully colored insects that compose the taxonomic family Lycaenidae. Nabokov had earlier held a job working on Harvard's butterfly collection, and the specimens he added are now celebrated as much by literary historians as by entomologists. Unaware of his literary talent, I visited him only to broaden my knowledge of butterflies and other insects, as I did all entomologists coming to the museum.

Nabokov entertained me with a personal story about butterflies that had an amusing twist. He'd met the captain of a ship that had recently visited Kerguelen, a remote archipelago off Antarctica seldom visited by anybody and with a still mostly unknown natural history. Nabokov was excited by the promise of such a remote environment. "Were there any butterflies?" he asked. "No," the captain responded. "None. Only a bunch of those little blue things."

Kerguelen! We rhapsodized about such places on Earth where unknown species awaited our discovery. The next year, with a three-year fellowship that allowed me to go anywhere and do anything I wished (provided I accomplished "something extraordinary" as the membership pledge demanded), I traveled around the world for unknown ant species: mountain forests in Cuba, the upper slopes of volcanic peaks in Mexico, rain forests in the South Pacific archipelagoes of New Caledonia and Vanuatu, the Saruwaged Range and other parts of New Guinea, and the mostly unexplored western fringe of the Nullabar Plain of Australia. In each place I searched joyously for something extraordinary. How could I help but find it in the adventure of natural history?

And so I understand very well both the Rocky Mountain elk hunter and the long-ago visiting lepidopterist. As Nabokov later expressed his vision: "The highest enjoyment of timelessness," he wrote, "is when I stand among rare butterflies and their food plants."

That is ecstasy, and behind the ecstasy is something else, which is hard to explain. It is like a momentary vacuum into which rushes all that I love. A sense of oneness with sun and stone. A thrill of gratitude to whom it may concern—to the contrapuntal genius of human fate or the tender ghosts humoring a lucky mortal.

For me, the same feeling returned fleetingly yet in full force in 2016, when I was invited to be the Honorary Principal Scientist (absentee, alas, due to age) on an

expedition to Heard Island. This second island off Antarctica is even more remote and rugged than Kerguelen
and features an active volcano. I wrote the expedition
leader Robert W. Schmieder that I was of course greatly
honored. I would accept, I added, on condition that the
team looked for ants. "You won't find any. It's just too
wet and cold. But just knowing you looked will make
me feel better."

15

GARDENS

Across the Paleolithic steppes of Europe and Asia, and back in time for more than 35,000 years, the ancestors of modern humans and their Neanderthal sister species buried their dead, leaving a treasure trove for present-day archeologists. In shallow graves, the ancients often included household items and upper body ornaments owned by the dead. A very few of the graves, only three per thousand years discovered thus far, had been elaborately decorated, suggesting a high rank of its occupant.

At the dawn of the Neolithic, characterized by agriculture, stored food surpluses, and settled villages, mourners began to add flower beds to the graves. The oldest known, dated to 13,700 years before the present, are at four sites in the Raqefet Cave at Mt. Carmel in northern Israel.

There is a logical reason flowers and gardens should rank so high on into modern times. The poet-naturalist Diane Ackerman has expressed it close to perfection:

A flower's fragrance declares to all the world that it is fertile, available, and desirable, its sex organs oozing with nectar. Its smell reminds us in vestigial ways of fertility, vigor, life-force, all the optimum expectancy and passionate bloom of youth. We inhale its ardent aroma and, no matter what our ages, we feel young and nubile in a world aflame with desire.

Flowers grace our literature, our fashions, our religious ceremonies. They announce our rites of passage and signify our celebrations. Made into boutonnières and garlands, they advertise our status, our purpose, and our demeanor for the day.

The beauty and scent of flowers did not emerge just for human delectation. Among the plants that produce them, namely the angiosperms (the more than 370,000 species of flowering plants of the world), their purpose is purely sexual. Flowers serve to attract pollinators, mostly insects, but in addition and according to species, a large array of birds and even a few mammals. The relationship is symbiotic, "mutually symbiotic" to be precise, in which both partners benefit from the relationship.

The strategies used by different species to achieve the symbiosis have evolved as though to delight the human artistic ego. Dendrolobium orchids, prominent tropical epiphytes, thrust clusters of delicate, modestly pink flowers from among their narrow leaves. Cyclamens offer seemingly amorphous red or white flowers in the midst of broad leaves, conspicuously bright with green and white, pretending to be the real stars of the show. Christmas coral columnea, challenging in brilliance any human art, dangle tubular red flowers from creeping vines amidst modestly small leaves.

These examples and countless others are plants arranging parts of their bodies for display. They depend on daytime animal partners that possess color vision. To equal degree, the relationship has evolved the other way. Some animals have evolved color vision by natural selection in order to harvest pollen and nectar offered by plants. The two kinds of partners, plant and animal, have coevolved as necessarily equal partners.

Humans have picked up on this worldwide symbiosis and taken it further, creating a new art form based on flowers. Floral design has a long and consistently brilliant history. In a 1640–45 imagined portrait of St. Francis, for example, the Spanish painter Francisco de Zubarán depicts the full length of the saint's robe emerging from a dark, sparse array of surrounding lilies that reflect the skin tones of the saint's upturned face. Shrouded by palm leaves, they repeat the angles of his robe.

Pursuing a wholly different mood, Jean Honoré Fragonard's *Mademoiselle Marie-Madeleine Guimard* expresses the self-indulgent excesses of the ancien régime on full display. The celebrated singer and dancer, judged to be one of the most beautiful in France by standards of the day, is surrounded by trains of her beloved roses that cross her head and run down her arms, while her form is wreathed by a cascade of other roses. Mlle. Guimard, in short, has become a flower.

Radically different from both of those classical examples is Paul Manship's stunning *The Moods of Morning* (1935). A sculpture of a supine man awakening is paralleled at the base by buds of many species, which push forward to an apical cluster of lilies and birds of paradise flowers. In the words of the art critic Victoria Jane Ream, the final bouquet is "more open, more intense, ending in an explosion of color and form" that represents the full awakening of the human subject.

In a different realm, flowers and fruits surrounded by sculpture have graced ornamental gardens since the beginning of civilization. Gardens in general have been the key instruments of agriculture, and agriculture was the means by which humanity evolved from Paleolithic hunting and gathering bands into civilized communities. Agriculture was invented multiple times independently worldwide, and on every inhabited continent except Australia, variously from about 12,000 years before the present in the Middle East to 5,000 years ago in the New World. It probably originated in one or the other of two ways. First was direct observation that edible seeds and fruit give rise to plants that produce more edible seeds and fruit. Alternatively, there would be a tendency for clusters of people to settle close to trees, shrubs, and herbaceous plants that are exceptionally productive, clearing the rest away. Domestication by selection of the most productive plants, hence agriculture, logically followed.

Gardens, like the natural world they represent, have a restorative and healing effect. Lay people have always felt their power, and scientists have tested and proven their salutary effects. In one such test, volunteers were shown first a stressful movie, then videos of either natural or urban settings. Those viewing scenes of nature experienced a subsidence of stress, measured in heartbeat rate, systolic blood pressure, facial muscle tension, and electrical skin conductance. But not those shown scenes of cities. Other studies have revealed, as expected, a similar reduction of stress symptoms in the presence of plants and aquariums prior to surgery and dental work. Furthermore, postsurgical patients recover more quickly, suffer fewer complications, and need smaller dosages of painkillers, when allowed to look out over natural environments or even just pictures of nature covering the hospital walls.

With the accumulation of wealth and the rise of social hierarchies, rulers and the wealthy could afford ornamental gardens for display and pleasure. Later, as for example in France, Spain, Japan, and Sri Lanka, these gardens were opened to the public. Today, public gardens are universal in most societies and give pleasure to all classes.

Homeowners in North America and most of Europe routinely convert their property into simulations of nature, usually consisting of lawns and mostly ornamental trees and shrubs. Yet, even though the benefits of nature are undeniable, landscape developers routinely commit two unintentional errors that harm the greater environment that they and their customers instinctively love. First is their obsession with lawns. People enjoy open spaces, and have most likely since the birth of humanity on Africa's savannas. A lawn, however, ranks among the worst possible environments worldwide, just above bare concrete. Every lawn (if we put aside the heroically rebellious dandelions) is a monoculture of alien species, ferocious consumers of water that also demand regular applications of fertilizers, toxic herbicides, and insecticides, the excess of which eventually seep into aquifers and stream headwaters.

Of equal concern, most landscape planners select ornamental trees and shrubs on the basis of their perceived beauty, with little or no concern for their geographical origin. Choosing exotic over native species has profound implications for the environment, most significantly the much higher production of insects supported by native plants. The wildlife ecologist Douglas W. Tallamy found that insects feeding on native woody plants in Pennsylvania produced four times the biomass of those feeding on exotics. Tallamy attributed the difference to the reduced ability of native insects with chewing mouthparts to eat alien plants, on which their species did not evolve. He found that native species supported thirty-five times more biomass of caterpillars in particular—chiefly moth larvae—than did exotic species.

Preferring trees and shrubs that produce *more* insects may seem the wrong way to go in landscape planning, but the opposite is the case. Insects, especially caterpillars, are the preferred diet of songbird nestlings. Native woody plants produce more birds and thereby an all-around richer ecosystem. A healthy population of birds provides a check on runaway population increases of insect pests, such as the Eurasian gypsy moth and emerald ash borer. Further, by supporting native bird populations at a sustainable level, we help ease the extinction risk of the rarest species among them. Replacing exotic plant species with alternative native species, which are comparably attractive and in every respect more interesting, is not just environmentally wise but also cost effective. It is personalized conservation at its best.

Biophilia, an instinct that is well supported by humanistic scholarship and scientific research, draws humans to the rest of life on Earth and provides us with deep satisfaction. While we have just begun to understand the evolutionary origin and neurobiological impact of this instinct, biophilic design has become a growing new development in architecture. The celebrated prototype is Frank Lloyd Wright's Fallingwater. Nestled tightly in woodland astride a waterfall, with a front facing downslope, its cantilevered two stories reminiscent of rock ledges and the murmuring of water heard from within, this dwelling, first occupied in 1937, still stands as an historic ideal of humanity's accommodation to nature.

A later biophilic masterwork is the Finnish Embassy in Washington, DC, designed by Mikko Heikkinen and Markku Komonen, and a recipient of major awards for design and energy efficiency. The interior is open all over to sunlight, bringing visibility in every direction. In my opinion, the building's most remarkable feature is its placement, not upon a manicured lawn but on the edge of undisturbed woodland in Rock Creek Park. Half of Finland Hall is enclosed by a glass wall at ground level that looks into the forest interior from only a few feet away.

By the end of the twenty-first century, if current demographic projections are correct and the planet is lucky, the human population will peak at roughly ten to twelve billion people, about 50 percent more than are alive today. The vast majority will live in cities. What will those cities be like, serving as the home of such an abnormally dense mammalian biomass? Will they become mountain ranges of stone, climate-controlled skyscrapers, fulfilling all dystopian visions, within which all the needs and desires of its inhabitants are met (much as for the macrotermites in their African termite hills)? Or will their planners find some way to let nature in, to bring people close to their own ancient genetic heritage? Will they enable us to remain fully human?

One means to bring nature in, which I proposed in *Half-Earth*, would be wall-sized television screens on which we could watch the best remaining wild habitats in our homes, in real time. My own personal choices would include a Serengeti waterhole, an Amazon forest canopy, an Indonesian coral reef, the journey of a tagged great white shark down the Pacific coast, and the microscopic bacterial and algal forest of a Yellowstone hot spring (thought by many experts to be a remnant of one of the earliest ecosystems on Earth).

It would furthermore not be too difficult to bring parts of the natural world into our cities, as argued by Timothy Beatley in *Biophilic Cities: Integrating Nature into Urban Design and Planning* (2011) and other like-minded urban architects. Existing parks and others being planned can be seeded to sustain a rich native biodiversity. Instinct and opportunity have cleared the way for a great metamorphosis. Vacant lots, interstices between buildings, and unused strips of riverside land can be allowed to regenerate their natural faunas and floras and used for recreation and education (as practiced by the Chicago Wilderness program, for example). Gardens and managed natural ecosystems can be planted on unused rooftop space and, using vines and natural cliffside vegetation, adorn even the sides of skyscrapers. There is no city whose inhabitants, once secure with personal necessities, could not enjoy all around their homes a flutter of butterflies, a parade of early spring flowers, and even an occasional thrilling swoop of a falcon through a small yet authentic stand of native forest.

The Ancient of Days, William Blake, 1794, his favorite work, shows the god Urizen creating the second Enlightenment. Urizen, thought Blake, was an evil god, because he invented science in order to force humanity into a single way of thinking.

V

The more closely we examine the properties of metaphors and archetypes, the more it becomes obvious that science and the humanities can be blended. In the borderland of new disciplines created, it should also be possible to reinvigorate philosophy and begin a new, more endurable Enlightenment.

16

METAPHORS

*B*ut look, the morn, in russet mantle clad, walks o'er the dew
of yon high eastward hill.

Horatio speaks thus to Marcellus in *Hamlet*, where a
simple exclamation "Look! Dawn has broken" would have
sufficed. But poetry we love and great poetry we cherish.
Poetry, and much of prose as well, is constructed with metaphor, defined by the literary critic I. A. Richards as "a shift,
a carrying over of a word from its normal use to a new use."

The invention of language in the first place, defined as
the expression of thought through sounds given arbitrary
meaning, was the supreme achievement of human evolution, genetic in its origin, cultural in its elaboration. Without the invention of language we would have remained
animals. Without metaphors we would still be savages.

Metaphors are the device by which new words, com-
binations of new words, and new meanings of words

are invented. An added poetic content invests language with emotion. Language impelled by emotion creates motivation, which drives civilization. The more advanced the civilization, the more elaborate its metaphors. Even the glossaries of physics and engineering are built with them.

A well-wrought phrase that implies the essential identity of the two things compared is a metaphor. Consider Yeats speaking of sisters who lived in the great English house Lisadell, in which

Pictures of the mind, recall
That gabble and the talk of youth,
Two girls in silk kimonos, both
Beautiful, one a gazelle.

In the metaphor, the critic Denis Donoghue observes, "The girl's nature goes over into the nature of the gazelle as if both came from one luminous source. That is what naming comes to: it is not a matter of glancing at an attribute here or there but of acknowledging a complete nature and giving it its destined name."

Metaphors are essential for humor as well. My two favorite examples are, first, for the reckless and domineering extrovert, "A bull in search of a china shop," and second, for the narcissicist, "A legend in his own mind."

Metaphors set the imagination free to search for vivi-fying images. They allow us to cross boundaries, deliver little shocks of aesthetic surprise and humor, and thereby achieve nuance and novel perspective. They permit an infinite expansion of language, and ideas identified by them. The growth has been exponential, with a doubling time of roughly three centuries. The number of words in Chaucer's time was close to 73,000, in Shakespeare's time 208,000, and in present-day use, according to the *Oxford English Dictionary*, 469,000. If technological and commercial jargon is included, the future number could easily double yet again and in a shorter space of time.

Words may be arbitrary in origin, but metaphors are not. Rather, they tend to fall into categories of innate human emotional response. Put differently, they are con-strained to some degree by instinct. Among animal-based metaphors, for example, "vulpine" (fox) means clever, secretive, selfish; "porcine" (pig) connotes fat, gluti-nous, untidy; "leonine" (lion) signifies the possession of strength, courage, majesty; serpentine (snake) means invidious, seductive, evil, and, in some societies, power-ful and beneficent. Across cultures, humans consistently use features of physical nature for metaphors. The sun, for example, represents enlightenment and wisdom; ice and snow, stillness, retreat, or death. The sea stands often for vastness, mother, birth, or mystery.

Metaphors are not intended to express the true nature of the entities that inspired them. Their meaning comes from the way a few of their traits affect our idiosyncratic human senses and emotion. In this perception, they are part instinctual and part learned, part genetic and part cultural. Predictable metaphors are woven together to create the archetypes of the creative arts. They are easily detected as stereotypical plots and characters in stories. They may be imprecise and even trite, but they are the bread and butter of literature and drama.

17

ARCHETYPES

W hereas metaphors are the building blocks of language, archetypes form part of the common groundwork of human emotion. Archetypes, composed of universal stories and images, have been recognized in Western culture since Aristotle analyzed Sophocles' tragic hero in *Oedipus Rex*. Their provenance is not an accident of cultural evolution. They are part of deep history, obedient to instinctive genetic biases acquired through evolution by natural selection. Some of their ultimate causes date back tens of thousands of years to a time when humans were spreading out from Africa to all the habitable globe. Others were forged in our distant animal ancestors, millions of years ago.

Examples of archetypes risen from genetically based instinct can be found in the propensity of present-day humans to acquire unreasoning phobias against snakes, spiders, and other ancient perils. It is also evidently displayed in the ideal habitat described by the savanna hypothesis, the open, elevated terrain next to a body of water. Both propensities have a demonstrable hereditary basis. In other words, they are instincts comparable to the hardwired habitat-seeking urge that drives all other organisms.

It makes sense that narratives in literature and the dramatic arts tend to cluster into natural categories. The propensity is clearly delineated in the cinema. Movies are art, and great movies are great art. If that much be granted, it is natural to ask which among them are the greatest, and, more importantly, why? In 1999 Richard D. McCracken, an anthropologist and cinema expert, polled members of the Directors Guild of America for their opinions of the ten most distinguished movies and scenes of history to that date, and also the reasons for their choice. The directors were asked to evaluate as relevant the quality of their favored story in the originality of its content, the performance of the actors, the theme music, and photography. In drawing on McCracken's analysis, and adding a few amateur but heartfelt opinions of my own, I have identified key archetypes I believe they variously illustrate. I've also speculated on the forces of natural selection that guided their genetic evolution. (Any errors made in grouping and interpretation are my responsibility.)

Archetypes in the Great Films

THE HERO.

Usually he, but increasingly she, fights singly or else rallies others in the struggle against enemies of seemingly overwhelming power. A viscerally pleasing scenario that can be rationally interpreted as the instinctive product of endless prehuman and primitive human warfare.

Examples:

Alexander Nevsky (1938). Destroyer of the Teutonic knights in an epic battle on the frozen Neva River.

Alien (1979) and *Aliens* (1986). Ripley, played by Sigourney Weaver, the ultimate feminist warrior, defeats some of the most terrifying aliens ever to invade a Hollywood set.

Bad Day at Black Rock (1955). One-armed World War II veteran thrashes a racist bully.

Billy Jack (1971). Indian martial arts expert whips a whole group of racist bullies.

Casablanca (1942). In the end, a noble character shines through Bogart's bluster and cynicism.

Fitzcarraldo (1982). With great effort, the protagonist, helped by a native tribe, overcomes the supreme difficulty of hauling a large ship out of one Amazonian river over a foothill ridge and into another.

Gunga Din (1934). In India, a hero improbably rises

from among the common people to save his colonial masters.

Henry V (1944). At the Battle of Agincourt, the king's superbly inspiring speech is followed by victory over the better-armed French.

High Noon (1952). The showdown between Gary Cooper's character and a gang of desperadoes bent on revenge is widely considered the best of its kind, although I prefer the electrifying final duel between Henry Fonda's assassin and Charles Bronson's avenger in *Once Upon a Time in the West* (1969).

The Last of the Mohicans (1992). Tribe versus tribe, then both versus a third tribe, with Hawkeye the noble savage.

THE TRAGIC HERO.

A tribal leader, and sometimes the entire tribe, possesses high status and power. Because of a fatal flaw, however, the leader or tribe fails and is destroyed. Throughout both prehistory and history, the mighty have thus risen high only to fall.

Examples:

The Caine Mutiny (1954). A naval captain rigidly follows protocol into insanity, met finally with revolt by his crew.

Citizen Kane (1941). A searing indictment of overreaching material power.

Gallipoli (1981). Tribe versus tribe (British invade Turkey); a perfect expression of the futility of modern war.

The Godfather, I, II, and III (1972, 1974, 1990). The greatest crime series in movie history tracks tribalism from power to greater power, thence to betrayal, degradation, and death.

Lawrence of Arabia (1962). A global hero watches his dream evaporate at the hands of the diplomatic leaders of his own, European tribe.

Sunset Boulevard (1950). Drama, ego, and insanity at the end of Norma Desmond's brilliant career.

THE MONSTER.

Prehuman and human ancestors lived constantly in fear of large predators that stalked and ate them. Venomous snakes across almost all of humanity's geographical range are still able to kill. Examples:

The Birds (1963). "Nature" strikes back with a frightening army of feathered avengers.

Forbidden Planet (1956). Fighting off the attack of a giant mind-generated demon in an early science-fiction classic.

Frankenstein (1931). "It's alive!" No better horror film has ever been made.

Invasion of the Body Snatchers (1955, remake 1978). Nobody can trust anybody else as the aliens guide human against human to conquer Earth.

Jaws (1957). The real thing out there in the water, terrifying from start to finish.

King Kong (1933). The original, the one I first sat through as an awestruck kid, terrifying at the start, gentle as a kitten at the close. The 2005 remake was a special effects masterpiece.

Nosferatu (1921). First in the century-long culture of cherished vampire films.

Psycho (1960). Tony Perkins portrays almost every familial horror that insanity can produce including murder, incest, and necrophily.

Silence of the Lambs (1990). Another, equally chilling depiction of an insane killer.

THE QUEST.

The prehuman and human ancestors were hunter-gatherers, forced to search constantly for game and vegetable assets within their foraging range. Discoveries of rich new sources of water, animal herds, or vegetable food were life-giving for the tribe, and the source of many of their stories and legends.

Examples:

2001: A Space Odyssey (1968). The origin and evolution of humanity, from mystery to mystery during a flight in search of an alien civilization.

Chariots of Fire (1981). Religion, honor, and courage in the struggle for Olympic gold.

Indiana Jones and the Last Crusade (1989). At last! The Holy Grail has been found! However, the discovery

has catastrophic consequences, ending in its second
disappearance.

Raiders of the Lost Ark (1981). The Ark of the Covenant
also recovered! Also with catastrophic consequences
and disappearance.

Treasure of the Sierra Madre (1948). A journey into dangerous
territory for easy riches succeeds, but ends in murder.

THE PAIR BOND.

*Outside sex, two men, or two women, or occasionally a man
and a woman, join forces in a fight for freedom while pursued by
hostile forces. A parable of the power, down through the ages, of
heroic altruism and cooperation.*

Examples:

The African Queen (1951). The radically different characters played by Bogart and Hepburn develop a close
relationship in colonial Africa during World War I
while fleeing Germans downriver.

Butch Cassidy and the Sundance Kid (1969). Partnership
in crime pays, all the way to the bloody end.

The Deer Hunter (1978). The Russian roulette scene
is the ultimate film expression of bravery and
sacrifice—intimate, scarifying, and unforgettable.

Lethal Weapon (1989). Buddies with radically different
personalities combine to bring some very bad guys
to justice.

The Man Who Shot Liberty Valance (1962). An unlikely
   alliance between two romantic rivals to defeat a
   gunslinging thug.
Thelma and Louise (1991). Suspected of murder, trapped
   at the edge of a canyon, this winsome pair choose
   suicide by automobile.

OTHER WORLDS.

*Prehuman and human tribes have always lived in a constant search
for additional territory, by discovery and conquest. For many millennia following the African breakout over 60,000 years ago,
human populations spread into lands completely new to our species. Meeting at the boundaries, they warred or formed alliances.*
Examples:

Alien (1979). An intense atmospheric exploration of
   a newly discovered planet, and a super-efficient
   monster parasite waiting there. With *Aliens* and
   The Thing (2011, remake of the 1982 film by John
   Carpenter), among the best science fiction horror
   movies ever made.
E.T. the Extra-Terrestrial (1982). An expression of what
   friendly aliens (and maybe human tribes) could
   be like.
Close Encounters of the Third Kind (1977). A second masterpiece on what friendly aliens can be imagined.
The Man Who Would Be King (1975). Sean Connery's

character travels over distant mountains to achieve power, riches, and, ultimately, betrayal and death.

Raiders of the Lost Ark (1981). The opening of the Ark (by hapless Nazis) reveals the terrors of the supernatural, faith-ordained world.

The magnetic pull of the other-world archetype reaches its highest intensity in the best of literary science fiction and cinema. These works of art are increasingly attractive in both story and accuracy of scientific and instrumental detail.

Examples:

Science fiction movies have matured perhaps more than any other category of cinema, as illustrated by one of my favorite early films, *The Angry Red Planet* (1960). A spaceship has landed on the surface of Mars. The astronauts, likeable next-door-neighbor kind of Americans, all wait with excitement for humanity's first close look at another planet's surface. A wall clock labeled BULOVA tells them the time. They look together through a porthole into a red haze. One asks, "Is there any sign of life?" Another answers, "Nothing. Just a bunch of plants." The audience's interest is kept alive by the foreboding thought that the plants might eat the astronauts, which they do.

Science fiction films have evolved into technically possible—or at least conceivable—adventure stories, many with end-of-Earth cataclysms, exoplanet colonization scenarios, and scientist heroes, who save the day (and perhaps the world). They include *Interstellar* (2014), in which pioneers leave a dying Earth and travel through a wormhole to settle a habitable planet. *The Martian* (2015) is a brilliant account of how ingenuity and grit could allow a stranded astronaut to survive on Mars until a rescue mission arrives. In *Europa Report* (2013), to conclude this sample of my personal favorites, the search for extraterrestrial aquatic organisms is conducted, successfully, in the icy sea of Jupiter's moon.

A superlative parallel effort in literary science fiction is Neil Stephenson's *Seveneves* (2015), praised by scientists for its technical fidelity, in which the moon blows up and its fragments head slowly for Earth, allowing refuges to be built to save the human species. If such ever happened, we can comfortably imagine it would be like this.

Overall, technical knowledge and fictive genius combined can provide unlimited enduring material for a blend of science and the creative arts.

The history of drama and its critical analysis, from the Greeks forward, is to a large degree the addition of language and culture to the hereditary, emotion-based repertories of an Old World primate. What I suggest here is that the instinctive residue has worked on language and culture as expressed in drama by clustering certain themes that evoke archetypes. In the modern era, the phenomenon appears in sharp outline in the history of motion pictures. This hypothesis is made plausible by the obvious imprint of ancestral animal traits throughout the anatomy and physiology of our bodies. This being the case, why is it then not also logical to include these special properties of human creativity as well? That question can also be phrased in two open questions, which I have tried to answer on earlier pages and which invite investigation by studies that combine science and the humanities: What is human nature?, and Why is there a human nature in the first place?

THE MOST DISTANT ISLAND

L et me illustrate the promise of a unified creativity in science and the humanities with two stories. The first concerns the most remote island in the world outside the Antarctic continent.

"Skerry" and "seastack" are words that are only rarely encountered in ordinary English speech. Of Norse origin, skerry means a small rocky island, and a seastack is a column of rock cut out by waves that have worn away the deposits of higher land around it. The two words have an almost magical pull on me. They summon my traveler's mind to mysterious forgotten places, offering sanctuaries somewhere on the other side of the ocean, constantly wave-lashed, worlds unto themselves. Even if the islands are too small to support human life, I want to go there!

I confess to being a neseophile, an inordinate lover of islands. My mind drifts easily to Isla Salas y Gómez, a skerry in the temperate Pacific off the coast of Chile. Standing alone in the open sea, measuring fifteen hectares in area and seven hundred seventy meters in maximum length, it is one of the smallest oceanic islands in the world. It began as a seamount, one of the thousands of volcanic peaks that rise from the ocean floor to varying depths beneath the surface. Isla Salas y Gómez is one of the rare peaks in the southeastern Pacific that has risen free to create a lonesome skerry. This relatively microscopic spot of land is also the most isolated island in the world. Its nearest neighbor is Easter Island, the planet's most isolated inhabited island, two hundred kilometers away.

Now, being a biologist with a special interest in geography, I especially wanted to know what kinds of plants and animals might have reached and now survive on this almost extraterrestrial place. The answer, I found, has been obtained by a very few visitors able to make it to Isla Salas y Gómez. This skerry is home to four land-dwelling species of plants, including a species of *Asplenium* fern also found on the remote Atlantic islands of St. Helena and Ascension. A very few insects are also present, feeding on foliage (I have not been able to find their names). A total of twelve species of seabirds go there to breed and raise their young. Of other terrestrial animals larger than insects there are none.

If we exclude the dozen or so species of seabirds as

visitors but include the still-uncounted insects and tiny nematodes and rotifers probably also present, the number of land-dwelling plant and animal species on this tiny lost world is probably under one hundred. A comparable area of tropical rain forest would possess a hundred times, even a thousand times, this number. Given that the island has existed for at least several centuries (Easter Islanders visited it when they still used catamarans for oceanic travel), why haven't more kinds of plants and animals colonized Isla Salas y Gómez? Why does it remain so close to lifeless?

The probable answer, biologists can now tell you, is that the diminished fauna and flora are in equilibrium. The great distances from the nearest other land allows only very few species to reach the skerry during long intervals of time, while the small size of Isla Salas y Gómez raises the rate at which they disappear. When the immigration rate equals the extinction rate, the number of species present at any given time remains very low. As a result, Isla Salas y Gómez adds another distinction to its "most modest" list: smallest flora and fauna of any island outside the world's polar regions.

Isla Salas y Gómez cannot support people. If it never had been seen, could it be real? The question is not as nonsensical as it must at first seem. It is just a version of the falling tree paradox: if there is no one in the forest to hear, does the fall of a tree make a sound? Common sense says that the answer is immediately obvious. The tree cannot fall without sending forth a wave of compressed air.

But a "sound" that has meaning for our species requires a human to hear that change in the air. A physicist and a biologist together may predict and simulate in fine detail the first crack of the trunk, the ominous susurration of the downward arcing canopy falling, the snaps and crashes of the plummeting branches, and the final *thump!* of the trunk hitting the ground. But neither the scientists nor anyone else can hear the actual fall. A human or recording device on the scene is necessary. Otherwise the event has no meaning. Nietzsche captured the larger point when he had Zarathustra address the Sun: "You great star! What would your happiness be if you had not those for whom you shine!"

Wallace Stevens developed more fully the import of the paradox in his 1943 poem "Somnambulisma" (that is, sleepwalking), using the image of an unseen ocean and unheard surf.

> *On an old shore, the vulgar ocean rolls*
> *Noiselessly, noiselessly, resembling a thin bird,*
> *That thinks of settling, yet never settles, on a nest.*
>
> *The wings keep spreading and yet are never wings.*
> *The claws keep scratching on the shale, the shallow shale,*
> *The sounding shallow, until by water washed away.*

The generations of the bird are all
By water washed away. They follow after,
They follow, follow, follow, in water washed away.

Without this bird that never settles, without
Its generations that follow in their universe,
The ocean, falling and falling on the hollow shore,

Would be a geography of the dead; not of that land
To which they may have gone, but of the place in which
They lived, in which they lacked a pervasive being,

In which no scholar, separately dwelling,
Poured forth the fine fins, the gawky beaks, the personalia,
Which, as a man feeling everything, were his.

The critic Helen Vendler broadens the key question as well as can be phrased: "If there did not exist, floating over us, all the symbolic representations that art and music, religion, philosophy, and history, have invented, and afterward all the interpretations and explanations of them that scholarly activity have passed on, what sort of people would we be?"

Neither the question nor the answer is rhetorical. There would be no literature, little or no abstract or symbolic language, no tribal government (with a radius greater than can be run on foot in a day). The technology would be Paleolithic, and art would still be crude figurines and stick figures drawn on rock walls, with little meaning left to decipher. Science and technology would consist of the sharpening of spear points, the knapping of stone axes, and perhaps the piercing of snail shells to thread for necklaces.

IRONY: A VICTORY OF THE MIND

The theme of humanistic science converging on the scientific humanities is provided a commanding example in the haunting ballad "Send in the Clowns," Stephen Sondheim's signature piece in the Broadway production *A Little Night Music*. Written in 1973 with the actress Glynis Johns in mind, "Send in the Clowns" has been performed by a host of leading artists. Its rendition by Judy Collins was named Song of the Year at the Grammy Awards ceremony of 1976.

The central character and voice of the play is Desirée, a beautiful and successful actress, who many years earlier had an affair and bore a child with Fredrik, a lawyer. This charming fellow, even while unaware he is a father, proposes marriage to Desirée. Wedded to her career, she refuses. When the action commences, Fredrik is trapped in a failed, unconsummated marriage with an equally beautiful and much younger woman. This time Fredrik is the one who refuses. An embittered Desirée responds with an ironic little song that begins,

Isn't it rich? Are we a pair?
Me here at last on the ground, you in mid-air.
Send in the clowns.

Desirée, we assume, must feel defeated, disappointed, and, indeed, furious inside. The many performing artists on stage and off have varied greatly in the way they express Desirée's emotions. The critics of their renditions, while almost uniformly admiring, nevertheless tend to be puzzled by the clowns. They made more sense when Sondheim explained the song as one of regret and anger. To send in clowns is a theatrical expression. It means that if things are going badly, "Let's do the jokes."

Regret and anger are surely expected. Who would not feel them—especially a celebrated actress like Desirée? But I find more weight in the stoic lyrics of this waltzy little piece. When read through straight and with consistent expectation, it is a textbook example of pure irony. The emotions it expresses date back to the origin of literacy, in the Greek *eirōneia*, broadly meaning dissimulation. It evolved as an emotional trait, rhetorical in nature, unique to humanity.

Irony is a device in speech and literature in which the

properties of a process or an entity are described by their exact opposite. It includes the Emperor's new clothes, the thunderous silence, the Biggest Little City, the calm peacefulness of a coiled rattlesnake, the "touch up" of an infantry battle, and, even in so august a temple of astrophysics where irony is never intended, the coexistence of the endless universe we can see within the equally endless existence of parallel universes. Irony creates a new level of meaning. It amuses, emphasizes, and softens the brutality of real life.

Few of the performing artists of "Send in the Clowns" whose recordings I have heard express these qualities as I feel they should. In varying degree, the artists translate the emotions into what they perceive behind the irony, and without forcing themselves out of their customary personal styles. Judy Collins is warm, sweet, and loving. Of all the singers, she is the partner you would most regret leaving. Barbra Streisand delivers her own muscular musicality, her voice and mood rising and falling dramatically. Glynis Johns is stricken with disbelief and anger, while Judi Dench is just grief-stricken. Carol Burnett is the very image of disappointed stoicism. Catherine Zeta-Jones embellishes her delivery with facial and tonal drama. Sarah Vaughan infuses it with nuances of a love song and a touch of jazz. Frank Sinatra is the wrong gender.

Glenn Close alone gets it right—start to finish. She is, in my admittedly personal judgment, perfect. Her demeanor at the microphone, her calm erect posture, her ironic smile suggest a cultivated and highly intelligent woman on the edge of middle age. Her Desirée is deeply disappointed, angry, and perhaps resigned, but still open to the distant possibility that the world might yet change her way. The importance of the situation is recognized in subtle mannerisms. The word "clowns" is pronounced with care and slight emphasis, each letter enunciated clearly. Desirée is a walking tragedy, but her demeanor is entirely civilized, and the more powerful for it.

Don't you love farce? My fault, I fear.
I thought that you'd want what I want,
Sorry my dear.
But where are the clowns? Send in the clowns.
Don't bother, they're here.

Anger, jealousy, and retribution are animal emotions. They are part of the instinctual programs already set in place in the hypothalamus and other emotional control centers of our ancestors tens of millions of years ago. Irony is something different. It is ours alone, cerebral, pacific, and shaped substantially by cultural evolution in social environments created by language. To explain animal emotions we must lean toward biology, engaging the humanities of course. To explain irony requires the reverse.

THE THIRD ENLIGHTENMENT

Contrary to common belief, the humanities are not distinct from science. No fundamental chasm in the real world or process of the human mind separates them. Each permeates the other. No matter how distant phenomena addressed by the scientific method may seem from ordinary experience, no matter how vast in expanse or microscopic its purview, all scientific knowledge must be processed by the human mind. The act of discovery is completely a human story. Its telling is a human achievement. Scientific knowledge is the idiosyncratic, absolutely humanistic product of the human brain.

It follows that the relationship between science and the humanities is fully reciprocal. Regardless how subtle, fleeting, and personalized human thought may be, all of it has a physical basis ultimately explainable by the scientific method.

If science is thereby the bedrock of the humanities, the humanities have the farther reach. Where scientific observation addresses all phenomena existing in the real world, scientific experimentation addresses all possible real worlds, and scientific theory addresses all conceivable real worlds, the humanities encompass all three of these levels and one more, the infinity of all fantasy worlds.

The European Enlightenment that progressed from the seventeenth into the late eighteenth centuries divided knowledge into the three great branches of learning: the natural sciences, the social sciences, and the humanities. The social sciences have now largely split amoeba-like into two divisions, one fusing with the natural sciences and the other hewing closely in language and style to the humanities. Contributions from the social sciences of the first kind can be seen in premier journals such as *Nature*, *Science*, and *The Proceedings of the National Academy of Sciences*. Those of the second kind, allegiant to the humanities, are to be found in *The New Yorker*, *The New York Review of Books*, *Public Interest*, and *Daedalus*, the journal of the American Academy of Arts and Sciences.

Science and the humanities may still remain apart, but they are ever more closely bonded in many ways. The degree of their relationship forms a continuum. At the extreme scientific end, the style of a typical research report in a professionally respected journal is relentlessly factual in content, heavy with observation and analysis, ostentatiously cautious in conclusions, and dull in the reading. Speculation, if any is ventured at all, must be presented as a hypothesis subject to new observations and experimentation. Metaphors, which are regarded by professionals as the equivalent of a lighted match in an ammunition dump, must be sparse and handled gingerly.

At the opposite end of the science–humanities continuum, where dwell the most creative of the creative arts, metaphors are coin of the realm. The emotional bump their aesthetic surprises deliver, whether in literature, music, or the visual arts, is the goal of the artist's effort, and novelty and virtuosity are its measure. The cognoscenti tend to speak about a scientific discovery in detail but not about the scientist, and, in reverse, critics in the creative arts say a great deal about the artist but not much about the art.

The blending of scientific and humanistic thought is increasing with time. The seismic gap that once separated them, made famous in 1954 by C. P. Snow's "two cultures," has closed, not by a narrow bridge but by a broad borderland of emerging new disciplines.

As science and the humanities draw closer together, the synergy between them is accelerating. The humanities have always been viewed as an ensemble of disciplines that explain "what it means to be human." However, this is not quite correct. They have described a goodly part of the human condition, but largely failed to explain what it all means. To achieve this goal will require a great deal more of the information available from scientific research than has been used by scholars of the humanities.

A notable characteristic of poets and other creative artists judged of major stature, along with the best of critics who assess their work, is ignorance of the very biology they celebrate. They are astonished when confronted by the architecture of the human body, charted from organ to molecule; by the true range of the human senses; by the turbulent and always chancy evolutionary history of the hominins; and not least by the living world that gave us birth and upon which our every breath depends. But as a rule they remain confidently ignorant of the details. They are content to talk exclusively to one another.

What exactly have we really learned, for example, from the immense library of novels that the public devours? T. S. Eliot's assessment is hard to refute: "Knowledge of life obtained through fiction is only possible by another stage of self-consciousness. That is to say, it can only be a knowledge of other people's knowledge of life, not of life itself."

A salient emotional trait of human nature is to watch fellow humans closely, to learn their stories, and thereby to judge their character and dependability. And so it has ever been since the Pleistocene. The first bands classifiable to the genus *Homo* and their descendants were hunter-gatherers. Like the Kalahari Ju/'hoansi of today, they almost certainly depended on sophisticated cooperative behavior just to survive from one day to the next. That, in turn, required exact knowledge of the personal history and accomplishments of each of their groupmates, and equally they needed an empathetic sense of the feelings and propensities of others. It gives deep satisfaction—call it, if you will, a human instinct—not just to learn but also to share emotions stirred by the stories told by our companions. The whole of these performances pays off in survival and reproduction. Gossip and storytelling are Darwinian phenomena.

A major cause of the alarming decline in public esteem and support of the humanities is their overly narrow focus on the human condition during present and recent historical times. As a formal definition of the humanities, that might at first seem acceptable. But it has confined humanistic awareness almost entirely to the tiny bubble within the vast physical and biological world in which our species originated and in which we continue to exist. Another effect of this narrow epistemological focus is that it renders the humanities rootless. Although humanistic arts and analyses superbly capture details of history, they remain largely unaware and uncaring about the evolutionary events of prehistory that created the human mind, which after all created the history on which the humanities focus. Further, the creative arts and critical analyses leave unknown and unmentioned most of the nonhuman physical and biological processes that churn ceaselessly around and through each of us. We remain largely blind to the environment and the forces within it that guide us to whatever destiny our activities foreordain.

In their own way, scientists are equally unprepared for collaboration with creative artists and scholars of the humanities. The great majority of scientists are journeymen, researchers living out their careers within small specialized territories of knowledge and inquiry (nowadays often called silos). They can tell you almost everything known about, say, cell membranes, or the megalomorph spiders, or whatever other narrow subject in which they are experts, but not much in depth about anything else. The reason is that to be a real scientist—and not, regardless of how gifted, a journalist, a nonfiction writer, or a historian of science—is to make a certifiable scientific discovery. The acid test among professionals is to be able to finish this sentence: "I discovered . . ." The importance of the discovery is judged by peers who dwell in the same and nearby silos. Real scientists seek above all the acceptances and esteem of their peers; public approbation is secondary. They tend to value election to a national academy of science over a prize given to a best-selling book.

The necessarily rigorous definition of authentic science is the reason that the great majority of scientists are content to be journeymen. To conduct original scientific research requires an apprenticeship, during which at the beginning broad subjects are learned, techniques mastered, and finally, in most cases, postdoctoral research conducted in collaboration with a senior scientist or team of scientists. The candidate chooses his or her specialization by personal interest and opportunity. In biology, which is the discipline closest in life and mind to the humanities, the young scientist develops a range of learning and experience that has been aptly called "a feel for the organism."

Because most scientific knowledge is growing exponentially, doubling every ten to twenty years according to subject, specialties within disciplines multiply—while at the same time narrowing in scope. In the early 1950s, when I was a graduate student, an original research report in biology was typically authored by one to three contributors. The historic 1953 article in *Nature* by James D. Watson and Francis Crick, first to describe the structure of DNA, was typical of opportunities that lay open to scientists working either alone or in very small teams. Today much larger teams are the rule. Dozens of coauthors heading a single article are not unusual. In a few subjects, such as the description of the complete DNA of an important species, their number may exceed a hundred.

The 1950s and 1960s were the heroic age of modern biology, when a small number of talked-about champions made impressive advances against long odds. The excitement they generated was reflected in the popular culture. In the 1953 original film version of *War of the Worlds*, an alien spaceship enclosed in a large meteorite has just struck Earth—in southern California, of course. As a crowd of locals stand around the crater, wondering what it means, Ann Robinson's character, a young local schoolteacher with that unbearably feminine sweetness popular in the 1950s, says to Gene Barry's character, a visitor, "There's a scientist coming from Pacific Tech. He'll tell us." If the same scene were repeated today, the line would have to be, "There are teams coming from NASA and Cal Tech. They'll try to figure out what's going on."

Just as in the humanities, necessary specialization has driven biologists and other scientists into territories growing ever smaller. Much of the advanced technology and technical language of one discipline is at best only partially comprehensible to experts in other disciplines, even those closely related.

Will there be heroic ages of the intellect in the future? I feel certain there will be, and especially in the new borderland disciplines that combine scientific discovery with the innovations and insights of the humanities. It will come in multiple dimensions. Outside our sensory bubble are countless prospects for the creative arts. The challenge is to translate the previously unperceived into the limited audiovisual world of human consciousness. The advance will also come from an increasingly clear view of the biological origins of human consciousness, which can bring prehistory into alignment with history. And, finally, the advance will come with understanding the evolutionary forge in which culture was shaped, slowly and often painfully, from animal instinct.

The philosopher's stone of human self-understanding is the relation between biological and cultural evolution. Why are human beings built and behaving in such and such a way and not some other? Only now, two-and-a-half millennia after the Athenian agora, are we beginning to understand the reason why some of our social behavior is hardwired instinct, some of it is acquired by genetically predisposed learning, and the rest is the product of cultural invention. The whole of it can be deeply understood only as a stage in our long-term evolution and not just through descriptions of the contemporary human condition.

Meanwhile, it needs to be recognized, and talked about more frankly, that for philosophy the elephant in the kitchen is organized religion. More precisely, the understanding of the human condition often foretold by the blending of science and religion is inhibited by the intervention of supernatural creation stories, each defining a separate tribe. It is one thing to hold and share the elevated spiritual values of theological religion, with a belief in the divine and trust in the existence of an afterlife. It is another thing entirely to adopt a particular supernatural creation story. Faith in a creation story gives comforting membership in a tribe. But it bears stressing that not all creation stories can be true, no two can be true, and most assuredly, all are false. Each is sustained by blind tribalistic faith alone.

The study of religion is an essential part of the humanities. It should nonetheless be studied as an element of human nature, and the evolution thereof, and not, in the manner of Christian bible colleges and Islamic madrassas, a manual for the promotion of a faith defined by a particular creation story.

Meanwhile, because humanity is still swept along by animal passions in a digitalized global world, and because we are conflicted between what we are and what we wish to become, and because we are drowning in information and starved for wisdom, it would seem appropriate to return philosophy to its once esteemed position, this time as the center of a humanistic science and a scientific humanities.

How might such a restoration be accomplished? It is worth bearing in mind that philosophy flourished within Western civilization during two bursts of creativity that lasted about 150 years each. Their essence has been concisely described by Anthony Gottlieb in *The Dream of Enlightenment*, a history of the rise of modern philosophy.

> The first was the Athens of Socrates, Plato, and Aristotle, from the middle of the fifth century to the late fourth century BC. The second was in northern Europe, in the wake of Europe's wars of religion and the rise of Galilean science. It stretches from the 1630s to the eve of the French Revolution in the late eighteenth century. In those relatively few years, Descartes, Hobbes, Spinoza, Locke, Leibniz, Hume, Rousseau, and Voltaire—most, that is, of the best-known modern philosophers—made their mark.

The second efflorescence of philosophy, hence the foundation of the second Enlightenment, had mostly faded by the early 1800s, when science failed to meet its own grandiose expectations, and the humanities alone could not pick up the extra burden. Today, what passes for twenty-first-century philosophy is to a large extent punditry, in which commentaries on current issues are issued mostly from scholars trained in the humanities and economics. The real limitation of present-day philosophy is not clashes of authorial logic but incoherence, due chiefly to inattention to science. This is curious, since we are in what can reasonably be called the Age of Science, and science is positioned to combine with the humanities to rekindle the spirit of the earlier Enlightenments. I believe that the two, meeting in common inquiry, can at last solve the great questions of philosophy. It is time to ask again, forthrightly and with a great deal more confidence than ever before, the great questions of history.

At the base, we need to explore ever more deeply the meaning of humanity, why we exist as opposed to have never existed. And further, why nothing even remotely like us existed on Earth before. The grail to be sought is the nature of consciousness, and how it originated. Equally fundamental is the origin and proliferation of life as a whole.

In a narrower focus, how are we to explain the existence of two genders—or more, considering the range of sexuality acknowledged today? Why is there sex in the first place? If we could reproduce parthenogenetically, or just bud offspring from our bodies, life would be so much simpler. These are not idle questions, to be left to salonists and after-dinner guests. They are not mind games, nor exercises to sharpen skills in logic. They address matters literally of life and death.

We must ask not just how, but why, we must die of old age if nothing else, and moreover why we are fixed by an exact schedule of genetically programmed growth and decline. And now, finding ourselves in the Age of Artificial Intelligence, we are forced to define, with exactitude, a human being. Might we build one from chemicals taken off the shelf, at least to the level of a fertilized egg begun with a designated genome? And if that proves impossible, which I doubt, we need to seriously discuss humanoid robots, endowed with emotions and even the power of creativity.

Scientists and scholars in the humanities, working together, will, I believe, serve as the leaders of a new philosophy, one that blends the best and most relevant from these two great branches of learning. Their effort will be the third Enlightenment. Unlike the first two, this one might well endure. If so, it will bring our species closer to realizing the prayer for reason inscribed by Diogenes and still visible in original form on the Oinoanda stoa in the ancient Greek region of Lycia.

Not least for those who are called foreigners, for they are not foreigners. For, while the various segments of the Earth give different people a different country, the whole compass of this world gives all people a single country, the entire Earth, and a single home, the world.

REFERENCES AND
FURTHER READING

CHAPTER 1: THE REACH OF CREATIVITY

Bly, Adam, ed. *Science is Culture: Conversations at the New Intersection of Science + Society.* New York: Harper Perennial, 2010.

Boorstin, Daniel J. *The Discoverers.* New York: Random House, 1983.

Carroll, Joseph, Dan P. McAdams, and Edward O. Wilson, eds. *Darwin's Bridge: Uniting the Humanities and Sciences.* New York: Oxford University Press, 2016.

Greenblatt, Stephen. *The Swerve: How the World Became Modern.* New York: W. W. Norton, 2011.

Jones, Owen D., and Timothy H. Goldsmith. "Law and behavioral biology." *Columbia Law Review* 105, no. 2 (2005): 405–502.

Koestler, Arthur. *The Act of Creation.* London: Hutchinson and Co., 1964.

Pinker, Steven. *The Language Instinct: The New Science of Language and Mind.* New York: William Morrow, 1994.

Poldrack, Russell A., and Martha J. Farah. "Progress and challenges in probing the human brain." *Nature* 526, no. 7573 (2015): 371–379.

Ryan, Alan. *On Politics: A History of Political Thought, Book One: from Herodotus to Machiavelli; Book Two: from Hobbes to the Present.* New York: W. W. Norton, 2012.

Sachs, Jeffrey D. *The Price of Civilization: Reawakening American Virtue and Prosperity.* New York: Random House, 2011.

Watson, Peter. *Convergence: The Idea at the Heart of Science.* New York: Simon & Schuster, 2016.

Wilson, Timothy D., et al. "Just think: The challenges of the disengaged mind." *Science* 345, no. 6192 (2014): 75–77.

CHAPTER 2: THE BIRTH OF THE HUMANITIES

Altmann, Jeanne, and Philip Muruthi. "Differences in daily life between semiprovisioned and wild-feeding baboons." *American Journal of Primatology* 15, no. 3 (1988): 213–221.

Ball, Philip. *The Music Instinct: How Music Works and Why We Can't Do Without It.* New York: Oxford University Press, 2010.

Biesele, Megan, and Robert K. Hitchcock. *The Ju/'hoan San of Nyae Nyae and Namibian Independence: Development, Democracy, and Indigenous Voices in Southern Africa.* New York: Berghahn Books, 2011.

Cesare, Giuseppe Di, et al. "Expressing our internal states and understanding those of others." *Proceedings of the National Academy of Sciences USA* 112, no. 33 (2015): 10331–10335.

de Waal, Frans. *Chimpanzee Politics: Power and Sex Among Apes.* New York: Harper & Row, 1982.

de Waal, Frans. *The Age of Empathy: Nature's Lesson for a Kinder Society.* New York: Random House, 2009: p. 89.

Fox, Robin. *The Tribal Imagination: Civilization and the Savage Mind.* Cambridge, MA: Harvard University Press, 2011.

Gottschall, Jonathan. *The Rape of Troy: Evolution, Violence, and the World of Homer.* New York: Cambridge University Press, 2008.

Greenblatt, Stephen. *The Swerve: How the World Became Modern.* New York: W. W. Norton, 2011.

Hare, Brian, and Jingzhi Tan. "How much of our cooperative behavior is human?" In Frans B. M. de Waal and Pier Francesco Ferrari, eds., *The Primate Mind: Built to Connect with Other Minds.* Cambridge, MA: Harvard University Press, 2012, pp. 192–193.

Kramer, Adam D. I., Jamie E. Guillory, and Jeffrey T. Hancock. "Experimental evidence of massive-scale emotional contagion through social networks." *Proceedings of the National Academy of Sciences, USA* 111, no. 24 (2014): 8788–8790.

Krause, Bernie. *The Great Animal Orchestra: Finding the Origins of Music In the World's Wild Places.* New York: Little, Brown, 2012.

McGinn, Colin. *Philosophy of Language: The Classics Explained.* Cambridge, MA: MIT Press, 2015.

Patel, Aniruddh D. *Music, Language, and the Brain.* New York: Oxford University Press, 2008.

Patel, Aniruddh D. *Music and the Brain.* Chantilly, VA: The Great Courses, The Teaching Co., 2015.

Thomas, Elizabeth Marshall. *The Old Way: A Story of the First People.* New York: Farrar, Straus and Giroux, 2006.

Tomasello, Michael. *The Cultural Origins of Human Cognition.* Cambridge, MA: Harvard University Press, 1999.

Wiessner, Polly W. "Embers of society: Firelight talk among the Ju/'hoansi bushmen." *Proceedings of the National Academy of Sciences, USA* 111, no. 39 (2014): 14027–14035.

CHAPTER 3: LANGUAGE

Bickerton, Derek. *More than Nature Needs: Language, Mind, and Evolution*. Cambridge, MA: Harvard University Press, 2014.

Boyd, Brian. *On the Origin of Stories: Evolution, Cognition, and Fiction*. Cambridge, MA: Belknap Press of Harvard University Press, 2009.

Carroll, Joseph. *Literary Darwinism: Evolution, Human Nature, and Literature*. New York: Routledge, 2004.

Eibl-Eibesfeldt, Irenäus. *Human Ethology*. New York: Aldine de Gruyter, 1989.

Gottschall, Jonathan. *The Rape of Troy: Evolution, Violence, and the World of Homer*. New York: Cambridge University Press, 2008.

Lamm, Ehud. "What makes humans different." *BioScience* 64, no. 10 (2014): 946–952.

Murdoch, James. "Storytelling—both fiction and nonfiction, for good and for ill—will continue to define the world." *Time* 186, no. 27/28 (2015): 39.

Pinker, Steven. *The Language Instinct: The New Science of Language and Mind*. New York: William Morrow, 1994.

Swirski, Peter. *Of Literature and Knowledge: Explorations in Narrative Thought Experiments, Evolution, and Game Theory*. New York: Routledge, 2007.

Tomasello, Michael. *The Cultural Origins of Human Cognition*. Cambridge, MA: Harvard University Press, 1999.

Tomasello, Michael. *A Natural History of Human Thinking*. Cambridge, MA: Harvard University Press, 2014.

Wilson, E. O. *Naturalist*. Washington, DC: Island Press, 1994.

CHAPTER 4: INNOVATION

Baldassar, Anne, et al. *Matisse, Picasso*. Paris: Éditions de la Réunion des musées nationaux, 2002.

Libaw, William H. *Painting in a World Transformed: How Modern Art Reflects Our Conflicting Responses to Science and Change*. Jefferson, NC: McFarland, 2005.

Richardson, John. *A Life of Picasso, Vol. 3: The Triumphant Years, 1917–1932*. New York: Knopf, 2007.

CHAPTER 5: AESTHETIC SURPRISE

Burguette, Maria, and Lui Lam, eds. *Arts: A Science Matter*. Hackensack, NJ: World Scientific Publishing, 2011.

Butter, Charles M. *Crossing Cultural Borders: Universals in Art and Their Biological Roots*. Privately published: C. M. Butter, 2010.

Hughes, Robert. *The Shock of the New: The Hundred-Year History of Modern Art*. New York: Knopf, 1988.

Ornes, Stephen. "Science and culture: Of waves and wallpaper." *Proceedings of the National Academy of Sciences, USA* 112, no. 45 (2015): 13747–13748.

Powell, Eric A. "In search of a philosopher's stone." *Archaeology* 68, no. 4 (July/August 2015): 34–37.

Romero, Philip. *The Art Imperative: The Secret Power of Art*. Jerusalem: Ex Libris, 2010.

Rothenberg, David. *Survival of the Beautiful: Art, Science, and Evolution*. New York: Bloomsbury Press, 2011.

Shaw, Tamsin. "Nietzsche: 'The lightning fire'." Review of K. Michalski, *The Flame of Eternity: An Interpretation of*

Nietzsche's Thought. New York Review of Books 60, no. 16 (2013): 52–57.

Sussman, Rachel. *The Oldest Living Things in the World.* Chicago: University of Chicago Press, 2014.

Talasek, John D. "Science and culture: Data visualization nurtures an artistic movement." *Proceedings of the National Academy of Sciences, USA* 112 no. 8 (2015): 2295.

Vendler, Helen. *The Ocean, the Bird, and the Scholar: Essays on Poets and Poetry.* Cambridge, MA: Harvard University Press, 2015.

Wald, Chelsea. "Neuroscience: The aesthetic brain." *Nature* 526, no. 7572 (2015): S2–S3.

CHAPTER 6: LIMITATIONS OF THE HUMANITIES

Dehaene, Stanislas. *Consciousness and the Brain: Deciphering How the Brain Codes Our Thoughts.* New York: Viking, 2014.

de Waal, Frans. *Chimpanzee Politics: Power and Sex Among Apes.* New York: Harper & Row, 1982.

de Waal, Frans. *Are We Smart Enough to Know How Smart Animals Are?* New York: W. W. Norton, 2016.

de Waal, Frans B. M., and Pier Francesco Ferrari, eds. *The Primate Mind: Built to Connect with Other Minds.* Cambridge, MA: Harvard University Press, 2012.

Hare, Brian, and Vanessa Woods. *The Genius of Dogs: How Dogs Are Smarter Than You Think.* New York: Dutton, 2013.

Harpham, Geoffrey Galt. *The Humanities and the Dream of America.* Chicago: University of Chicago Press, 2011.

Krause, Bernie. *The Great Animal Orchestra: Finding the Origins of Music In the World's Wild Places.* New York: Little, Brown, 2012.

Safina, Carl. *Beyond Words: What Animals Think and Feel.* New York: Henry Holt, 2015.

Whitehead, Hal, and Luke Rendell. *The Cultural Lives of Whales and Dolphins.* Chicago: University of Chicago Press, 2015.

CHAPTER 7: THE YEARS OF NEGLECT

Burns, Ken, and Ernest J. Moniz. "On the arts and sciences." *Bulletin of the American Academy of Arts & Sciences* 67, no. 2 (2014): 11–21.

Birgeneau, Robert J., et al. "Public higher education and the private sector." *Bulletin of the American Academy of Arts & Sciences* 67, no. 3 (2014): 7–17.

Brodhead, Richard H, and John W. Rowe, eds. *The Heart of the Matter: The Humanities and Social Sciences for a Vibrant, Competitive, and Secure Nation.* Cambridge, MA: The American Academy of Arts & Sciences, 2013.

Gonch, William, and Michael Poliakoff. *A Crisis in Civic Education.* Washington, DC: American Council of Trustees and Alumni, 2016.

Pforzheimer, Carl H. III. "Humanities, education and social policy: The Commission on the Humanities and Social Sciences." *Bulletin of the American Academy of Arts & Sciences* 68, no. 2 (2015): 20–21.

Saller, Richard, et al. "The humanities in the digital age." *Bulletin of the American Academy of Arts & Sciences* 67, no. 3 (2014): 25–35.

CHAPTER 8: ULTIMATE CAUSES

Flannery, Kent, and Joyce Marcus. *The Creation of Inequality: How Our Prehistoric Ancestors Set the Stage for Monarchy, Slavery, and Empire*. Cambridge, MA: Harvard University Press, 2012.

Grant, Andrew. "Evolution may favor limited life span." *Science News* 188 no. 1 (2015): 6.

Guadagnini, Walter. *Matisse*. Edison, NJ: Chartwell Books, 2004.

Hughes, Robert. *The Shock of the New: The Hundred-Year History of Modern Art*. New York: Knopf, 1988.

Shackelford, George T. M., and Claire Frèches-Thory. *Gauguin, Tahiti*. Boston: Museum of Fine Arts Publications, 2004.

Westneat, David E., and Charles W. Fox, eds. *Evolutionary Behavioral Ecology*. New York: Oxford University Press, 2010.

Wilson, E. O. *Sociobiology: The New Synthesis*. Cambridge, MA: Belknap Press of Harvard University Press, 1975.

Wilson, Edward O. *On Human Nature*. Cambridge, MA: Harvard University Press, 1978.

Wilson, Edward O. *The Meaning of Human Existence*. New York: Liveright, 2014.

CHAPTER 9: BEDROCK

Gottschall, Jonathan, and David Sloan Wilson, eds. *The Literary Animal: Evolution and the Nature of Narrative*. Evanston, IL: Northwestern University Press, 2005.

Haidt, Jonathan. *The Happiness Hypothesis: Finding Modern Truth in Ancient Wisdom*. New York: Basic Books, 2006.

Pagel, Mark. "Genetics: The neighbourly nature of evolution."

Review of A. Wagner, *Arrival of the Fittest: Solving Evolution's Greatest Puzzle* subtitle changed to *How Nature Innovates* (New York: Current, an imprint of Penguin Books, 2015). *Nature* 514, no. 7520 (2014): 34.

Wilson, Edward O. *The Social Conquest of Earth*. New York: Liveright, 2012.

Wilson, Edward O. *The Meaning of Human Existence*. New York: Liveright, 2014.

CHAPTER 10: BREAKTHROUGH

Antón, Susan C., Richard Potts, and Leslie C. Aiello. "Evolution of early *Homo*: An integrated biological perspective." *Science* 345, no. 6192 (2014): 45.

Brown, Kyle S. et al. "An early and enduring advanced technology originating 71,000 years ago in South Africa." *Nature* 491, no. 7425 (2012): 590–493.

Fox, Robin. *The Tribal Imagination: Civilization and the Savage Mind*. Cambridge, MA: Harvard University Press, 2011.

Heinrich, Bernd. *Racing the Antelope: What Animals Can Teach Us About Running and Life*. New York: Cliff Street, 2001.

Marchant, Jo. "The Awakening." *Smithsonian* 46, no. 9 (2016): 80–95.

Wilson, Edward O. *The Social Conquest of Earth*. New York: Liveright, 2012.

Wilson, Edward O. *The Meaning of Human Existence*. New York: Liveright, 2014.

Wrangham, Richard. *Catching Fire: How Cooking Made Us Human*. New York: Basic Books, 2009.

CHAPTER II: GENETIC CULTURE

Butter, Charles M. *Crossing Cultural Borders: Universals in Art and Their Biological Roots*. Privately published: C. M. Butter, 2010.

Lumsden, Charles J., and Edward O. Wilson. *Genes, Mind, and Culture: The Coevolutionary Process*. Cambridge, MA: Harvard University Press, 1981.

van Anders, Sari M., Jeffrey Steiger, and Katharine L. Goldey. "Effects of gendered behavior on testosterone in women and men." *Proceedings of the National Academy of Sciences, USA* 112, no. 45 (2015): 13805–13810.

CHAPTER 12: HUMAN NATURE

Boardman, Jason D., Benjamin W. Domingue, and Jason M. Fletcher. "How social and genetic factors predict friendship networks." *Proceedings of the National Academy of Sciences, USA* 109, no. 43 (2012): 17377–17381.

Eibl-Eibesfeldt, Irenäus. *Human Ethology*. New York: Aldine de Gruyter, 1989.

Graziano, Michael S. A. *Consciousness and the Social Brain*. New York: Oxford University Press, 2013.

Haidt, Jonathan. *The Righteous Mind: Why Good People Are Divided by Politics and Religion*. New York: Pantheon Books, 2012.

Orians, Gordon H. *Snakes, Sunrises, and Shakespeare*. Chicago: University of Chicago Press, 2014.

Rychlowska, Magdalena, et al. Heterogeneity of long-history migration explains cultural differences in reports of emotional expressivity and the functions of smiles. *Proceedings of the National Academy of Sciences, USA* 112, no.19 (2015): E2429–E2436.

Sussman, Anne, and Justin B. Hollander. *Cognitive Architecture: Designing for How We Respond to the Built Environment.* New York: Routledge, 2015.

Wilson, Edward O. *On Human Nature.* Cambridge, MA: Harvard University Press, 1978.

Wilson, Edward O. *The Social Conquest of Earth.* New York: Liveright, 2012.

CHAPTER 13: WHY NATURE IS MOTHER

Beatley, Timothy. *Biophilic Cities: Integrating Nature Into Urban Design and Planning.* Washington, DC: Island Press, 2011.

McKibben, Bill, ed. *American Earth: Environmental Writing Since Thoreau.* New York: Literary Classics of the U.S. distributed by Penguin Putnam, 2008.

Moor, Robert. *On Trails.* New York: Simon & Schuster, 2016.

Orians, Gordon H. *Snakes, Sunrises, and Shakespeare.* Chicago: University of Chicago Press, 2014.

Williams, Florence. *The Nature Fix: How Being Outside Makes You Happier, and More Creative.* New York: W. W. Norton, 2016.

Wilson, Edward O. *The Future of Life.* New York: Alfred A. Knopf, 2002.

Wilson, Edward O. *Half-Earth: Our Planet's Fight for Life.* New York: Liveright, 2016.

CHAPTER 14: THE HUNTER'S TRANCE

Cox, Gerard H. *Blood On My Hands*. Indianapolis, IN: Dog Ear Publishing, 2013.

Essen, Carl von. *The Hunter's Trance: Nature, Spirit, & Ecology*. Great Barrington, MA: Lindisfarne Books, 2007.

CHAPTER 15: GARDENS

Beatley, Timothy. *Biophilic Cities: Integrating Nature Into Urban Design and Planning*. Washington, DC: Island Press, 2010.

Buchmann, Stephen. *The Reason for Flowers: Their History, Culture, Biology, and How They Change Our Lives*. New York: Scribner, 2015.

Dadvand, Payam, et al. "Green spaces and cognitive development in primary schoolchildren." *Proceedings of the National Academy of Sciences, USA* 112, no. 26 (2015): 7937–7942.

Kellert, Stephen R., Judith H. Heerwagen, and Martin L. Mador, eds. *Biophilic Design: The Theory, Science, and Practice of Bringing Buildings to Life*. Hoboken, NJ: Wiley, 2008.

Ream, Victoria Jane. *Art In Bloom*. Salt Lake City: Deseret Equity, 1997.

Tallamy, Douglas W. *Bringing Nature Home: How You Can Sustain Wildlife with Native Plants*. Portland, OR: Timber Press, 2009.

Wilson, Edward O. *Biophilia*. Cambridge, MA: Harvard University Press, 1984.

CHAPTER 16: METAPHORS

Donoghue, Denis. *Metaphor.* Cambridge, MA: Harvard University Press, 2014.

CHAPTER 17: ARCHETYPES

Boyd, Brian, Joseph Carroll, and Jonathan Gottschall, eds. *Evolution, Literature, and Film: A Reader.* New York: Columbia University Press, 2010.

Coxworth, James E., et al. "Grandmothering life histories and human pair bonding." *Proceedings of the National Academy of Sciences, USA* 112, no. 38 (2015): 11806–11811.

Heng, Kevin, and Joshua Winn. "The next great exoplanet hunt." *American Scientist* 103, no. 3 (2015): 196–203.

McCracken, Robert D. *Director's Choice: The Greatest Film Scenes of All Time and Why.* Las Vegas, NV: Marion St. Publishing, 1999.

CHAPTER 18: THE MOST DISTANT ISLAND

MacArthur, Robert H., and Edward O. Wilson. *The Theory of Island Biogeography.* Princeton, NJ: Princeton University Press, 1967.

Vendler, Helen. *The Ocean, the Bird, and the Scholar: Essays on Poets and Poetry.* Cambridge, MA: Harvard University Press, 2015.

CHAPTER 19: IRONY: A VICTORY OF THE MIND

Sondheim, Stephen. "Send in the Clowns" from *A Little Night Music,* music and lyrics by Stephen Sondheim. New York: Studio Duplicating Service, 446 West 44th Street, 1973.

CHAPTER 20: THE THIRD ENLIGHTENMENT

Ayala, Francisco J. "Cloning humans? Biological, ethical, and social considerations." *Proceedings of the National Academy of Sciences, USA* 112, no. 29 (2015): 8879–8886.

Catapano, Peter, and Simon Critchley, eds. *The Stone Reader: Modern Philosophy in 133 Arguments*. New York: Liveright, 2016.

Cofield, Calla. "Science and culture: High concept art and experiments." *Proceedings of the National Academy of Sciences, USA* 112, no. 10 (2015): 2921.

Dance, Amber. "Science and culture: Oppenheimer goes center stage." *Proceedings of the National Academy of Sciences, USA* 112, no. 24 (2015): 7335–7336.

Gottlieb, Anthony. *The Dream of Enlightenment: The Rise of Modern Philosophy*. New York: Liveright, 2016.

Johnson, Mark. *Morality for Humans: Ethical Understanding From the Perspective of Cognitive Science*. Chicago: The University of Chicago Press, 2014.

Ornes, Stephen. "Science and culture: Charting the history of Western art with math." *Proceedings of the National Academy of Sciences, USA* 112, no. 25 (2015): 7619–7620.

Ruse, Michael, ed. *Philosophy After Darwin: Classic and Contemporary Readings*. Princeton, NJ: Princeton University Press, 2009.

Sachs, Jeffrey D. *The Price of Civilization: Reawakening American Virtue and Prosperity*. New York: Random House, 2011.

Schich, Maximillian, et al. "A network framework of cultural history." *Science* 345, no. 6196 (2014): 558–562.

Simontin, Dean Keith. "After Einstein: Scientific genius is extinct." *Nature* 493, no. 7434 (2013): 602.

Tett, Gillian. *The Silo Effect: The Peril of Expertise and the Promise of Breaking Down Barriers*. New York: Simon & Schuster, 2015.

Watson, Peter. *Convergence: The Idea at the Heart of Science*. New York: Simon & Schuster, 2016.

Weber, Andreas. *Biology of Wonder: Aliveness, Feeling, and the Metamorphosis of Science*. Gabriola Island, BC: New Society Publishers, 2016.

ACKNOWLEDGMENTS

The present work was conceived and written with the help of many friends and colleagues expert in variously multiple subjects visited by the author. Two deserve special mention for their essential roles in bringing the whole to completion: my research and editorial assistant Kathleen M. Horton, and adviser and editor Robert Weil, director of the Liveright Publishing Corporation, a division of W. W. Norton & Company.

CREDITS

49   "A wolf left his lair one evening in fine spirits . . ." Benjamin Carlson. *The Wolf and His Shadow.* 2015. Ink on illustration board. 20 x 30 inches. © Benjamin Carlson. Illustration from 25 Fables: Aesop's Animals Illustrated, curated by Bronwyn Minton, associate curator of art and research, National Museum of Wildlife Art, Jackson, WY.

119 "During interactions with strangers, the presence of . . ." M. Rychlowska, Y. Miyamoto, D. Matsumoto, U. Hess, E. Gilboa-Schechtman, S. Kamble, and P. M. Niedenthal. "Heterogeneity of long-history migration explains cultural differences in reports of emotional expressivity and the functions of smiles." *Proceedings of the National Academy of Sciences of the United States of America*, 112 (2015), E2429–E2436.

130 "I don't really need people but people need me . . ." Julia Roberts. YouTube speech on behalf of Conservation International. Used by permission.

138 "I came across fresh tracks of several elk, including . . ." Carl François von Essen. *The Hunter's Trance: Nature, Spirit, & Ecology.* Great Barrington, MA: Lindisfarne Books, 2007. Used by permission.

144 "The *L. rupicapra* holotype was very likely collected at a high-elevation . . ." Robert W. Taylor. Quote. Used by permission.

150 "A flower's fragrance declares to all the world that it is fertile . . ." Brief excerpt from page 50 of *Cultivating Delight: A Natural History of My Garden* by Diane Ackerman. Copyright © 2001 by Diane Ackerman. Reprinted by permission of HarperCollins Publishers.

179 "On an old shore, the vulgar oceans roll . . ." "Somnambulisma," from *The Collected Poems of Wallace Stevens* by Wallace

INDEX

Page numbers in *italics* refer to illustrations.